Arch Lake Woman

Peopling of the Americas Publications

Published for the Center for the Study of the First Americans

General Editors: Michael R. Waters and Ted Goebel

Arch Lake Woman

Physical Anthropology and Geoarchaeology

Douglas W. Owsley,
Margaret A. Jodry, Thomas W. Stafford Jr.,
C. Vance Haynes Jr., and
Dennis J. Stanford

With

Richard L. Jantz, James M. Warnica,
Tsunehiko Hanihara, Joanne Dickenson,
Laura Bergstresser, Ian G. Macintyre, M. Amelia Logan,
and John L. Montgomery

TEXAS A&M UNIVERSITY PRESS

College Station

This paper meets the requirements of ANSI/NISO, Z39.48-1992 (Permanence of Paper).
Binding materials have been chosen for durability.

Library of Congress Cataloging-in-Publication Data

Arch Lake Woman : physical anthropology and geoarchaeology / Douglas W. Owsley
... [et al.]. — 1st ed.
p. cm.
"Peopling of the Americas publications"
Includes bibliographical references and index.
ISBN-13: 978-1-60344-208-4 (cloth : alk. paper)
ISBN-10: 1-60344-208-1 (cloth : alk. paper)
1. Arch Lake Woman.   2. Arch Lake Burial Site (N. M.)   3. Roosevelt County
(N. M.)—Antiquities.   4. Antiquities, Prehistoric—New Mexico—Roosevelt
County.   5. Paleo-Indians—New Mexico—Roosevelt County.   6. Indians of North
America—Anthropometry—New Mexico—Roosevelt County.   7. Excavations
(Archaeology)—New Mexico—Roosevelt County.   I. Owsley, Douglas W.
E78.N65A82   2010
978.9'32—dc22
2010011142

Contents

Illustrations

All color plates are in section following page xvi.

Tables

Preface

The Arch Lake human burial site is located in eastern New Mexico on the highest promontory on the south side of the now-dry Arch Lake basin near the Texas border. When discovered and excavated in 1967, a geologic context of considerable antiquity was recognized. In February 2000 an interdisciplinary team reexamined the osteology, geology, archaeology, and radiocarbon age of the burial. This Paleoamerican woman was lying supine with her left arm semi-flexed, legs extended, and head to the southeast. Associated cultural materials include nineteen talc beads found near the lower part of her neck, a probable bone tool lying over her ribs, and dense red pigment associated with a unifacial stone tool near her left elbow. AMS radiocarbon dating results indicate an uncalibrated age of 10,020 ± 50 RC yr BP.

The skeletal analysis included comparing cranial and dental measurements and tooth crown discrete traits to those of other ancient North American remains and recent Native Americans. The morphology of Arch Lake is unique among early American crania because it has a short, broad vault. In contrast, its large size, short face, and low orbits correspond with other early crania and differ markedly from those of more recent Native Americans. When compared to other early crania, the Arch Lake woman is most similar to the Gordon Creek woman and the Horn Shelter No. 2 adult male. Of the comparative series, their archaeological sites are closest to Arch Lake geographically. The dental traits from Arch Lake and the Horn Shelter No. 2 juvenile are not representative of recent Native Americans.

Cultural affiliation of the Arch Lake woman is uncertain. Archaeological analysis included comparing burial assemblages and practices of five primary interments with uncalibrated radiocarbon ages between 9500 RC yr. and 10,020 RC yr. Burial goods accompanying Arch Lake are thought to include personal belongings worn or used at the time of burial: a necklace and a possible pouch tied at the waist containing ocher and a heavily resharpened flake knife. This aspect resembles the Wilson-Leonard burial, where a woman was interred with a shark's tooth at her neck and a heavily used stone tool at her side. More extensive burial assemblages and ceremonial treatments characterize Gordon Creek and Horn Shelter No. 2, which, like Buhl, included relatively unworn tools and items that may have been manufactured specifically for burial. The extended burial position at Arch Lake is a marked departure from the flexed burials at Gordon Creek, Horn Shelter No. 2, and Wilson-Leonard. Burial practice at Arch Lake and Gordon Creek included the use of red ocher.

The Arch Lake skeleton was in the process of eroding out of a road bank when it was discovered. The original investigators extended great effort to ensure its protection so that future study would be possible. This report fulfills their longstanding objective and adds new information on the people and burial practices of ancient North America.

Acknowledgments

This book is dedicated to Dr. F. Earl Green, who contributed so much to Paleoamerican geoarchaeology, and the late Carolyn Rose, Conservator and Chair, Department of Anthropology, National Museum of Natural History (NMNH), for expert guidance on stabilization and treatment of the consolidated matrix and skeleton.

The burial was found by Gregg Moore, Cecil Clark, and Linda Clark. James M. Warnica recognized the significance of the find, invited Dr. Green to assist with its examination, and assisted Smithsonian researchers with the study. Dr. Green recorded the stratigraphic provenance, offered advice on the excavation protocol and recovery of the skeleton en bloc, and photographed the final stage of the excavation on May 23, 1967 (see figs. 3a, 3b, and 4). Transport was assisted by John Bradley, Cecil Clark, Charles Harrison, Gregg Moore, and James Warnica. The osteological examination was assisted by Karin Bruwelheide, David Hunt, and Rebecca Kardash of the NMNH. Adhering, solidified matrix was skillfully removed from the exterior of the cranium by Carolyn Rose and Derek Finholt. Figures 5–10 and 19 were photographed by Chip Clark of the NMNH; figure 13 was provided by C. Vance Haynes Jr.; and figure 14 was provided by Ian G. Macintyre and M. Amelia Logan. Figures 1, 2, 11a, 11b, and 12 were drafted by Marcia Bakry of the NMNH using diagrams provided by Haynes, except for figure 2, which was provided by Phillip H. Shelley. Bakry also illustrated the flake tool shown in figure 19 and figure 20. Skye Sellars assisted with the field work. Sandra Schlachtmeyer provided editorial assistance. We thank Mike Waters and Britt Bousman for their editorial suggestions. The following individuals, all of the NMNH, also provided assistance: Aleithea Williams helped finalize the manuscript; Laurie Burgess reviewed the section discussing beads; and Vicki Simon reviewed the references for accuracy.

Comparative osteological data were collected through the courtesy of Hamline University, the Nevada State Museum, University of Colorado Museum, National Museum of Natural History, and U.S. Forest Service, with special thanks given to James Bedwell, Linda Cordell, Amy Dansie, Deborah Confer, Susan Myster, Al Redder, Sue Struthers, Donald Tuohy, and Richard Wilshusen. Al Redder assisted during archaeological data collection on the Horn Shelter No. 2 burials.

Partial funding for Haynes was provided by the University of Arizona Regents Professor Research Fund. The Arizona-NSF AMS Lab dated samples from the site. Initial OSL and AMS radiocarbon measurements were made by Stephen Stokes and Robert Hedges of Oxford University. Topographic surveying and sedimentary analysis were provided by anthropology students under the direction of Phillip Shelley, Eastern New Mexico University. SEM microprobe and the X-ray diffraction analyses were completed at the University of Arizona by Gary Chandler and Wes Bilodeau, respectively. Travel support for the research team was provided by a travel grant to Dennis J. Stanford from the National Geographic Society Committee for Research and Exploration (NGS grant #6775–00). Eastern New Mexico University and the Blackwater Draw Museum and Site provided facilities, student assistance, and partial funding for the radiocarbon analysis. Field work completed by Thomas W. Stafford Jr. was partially funded by Forrest Fenn.

The burial site is on private property. When the Arch Lake burial was first discovered, the site was on land belonging to Bennie Taylor of Elida, New Mexico. The present owner of the land is Tom Davis of Portales, New Mexico.

Arch Lake Woman

Color Plates

FIGURE 3a. The Arch Lake burial looking to the northwest as exposed by the El Llano Archaeological Society.

FIGURE 3b. View of the Arch Lake burial looking to the northwest and showing the grave profile in gray sand.

FIGURE 4. Left lateral view of the Arch Lake burial showing the unifacial stone tool and associated red ocher along and below left humerus (1) and probable bone tool (2).

FIGURE 5. Superior view of the Arch Lake skeleton after additional exposure in the laboratory.

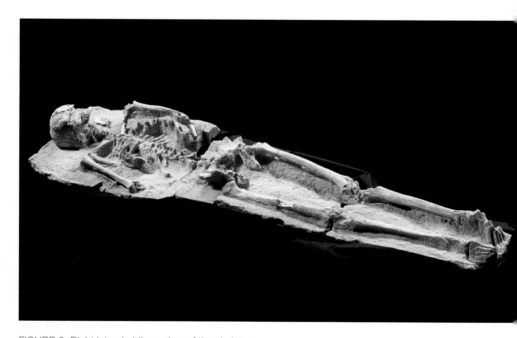

FIGURE 6. Right lateral oblique view of the skeleton.

FIGURE 10. Left lateral elbow region.

FIGURE 13. Photograph of the south wall stratigraphic column of the road cut.

FIGURE 14. Photomicrograph of sample showing a very poorly sorted ferruginous, calcareous silty sandstone.

FIGURE 19. Photograph of the unifacial tool and talc beads found with the burial.

Introduction

T HE Arch Lake human burial attracted little attention when it was discovered in 1967, even though the geologic context indicated considerable antiquity and it featured red pigment, a unifacial stone tool, a
probable bone tool, and nineteen beads. The Blackwater Draw Museum at
Eastern New Mexico University has curated the skeleton since its discovery
and removal. Here the burial was exhibited from 1969 to 1985, still partially
encased in surrounding matrix from within and below the burial pit.

In 1990 the Research Laboratory for Archaeology at Oxford University obtained a preliminary optically stimulated luminescence (OSL) age of 13,100 ±
2450 calendar years from soil from an overlying stratum. This, along with AMS
radiocarbon dating of bone and tooth organic carbon by the Oxford laboratory, provided the first chronometric confirmation of the skeleton's antiquity
(Stephen Stokes, pers. comm. 1991).

In February 2000 an interdisciplinary team organized by the Department
of Anthropology at the National Museum of Natural History reexamined the
burial to assess the following: its discovery history, geologic context, bone
chemistry, AMS radiocarbon age, skeletal and dental features, cultural affiliation, burial practice, and associated artifact assemblage. The results of this
work are reported herein. Arch Lake cranial and dental measurements and
discrete traits, as well as cultural aspects of burial practice, are compared to
other ancient North American remains, including those from Gordon Creek,
Horn Shelter No. 2, Wilson-Leonard, and Buhl (Breternitz et al. 1971; Green
et al. 1998; Guy 1998; Jantz and Owsley 1997, 2001, 2005; Muniz 2004; Redder
1985; Redder and Fox 1988; Steele 1998; Steele and Powell 1992, 1994; Sullivan
1998; Young 1988; Young et al. 1987).

Discovery and Excavation

The Arch Lake burial site is located on the highest promontory on the south side
of the now-dry Arch Lake basin in eastern New Mexico near the Texas border.

On May 20, 1967, avocational archaeologists Gregg Moore of Elida, New
Mexico, and Cecil Clark of Portales, New Mexico, discovered the grave while
examining a cut bank on the north side of a dirt road that bisected a late Pleistocene dune. They initially found the left half of a heavily carbonate-encrusted
mandible and subsequently located partially exposed and broken left parietal
fragments of the cranium approximately 0.6 meters south and down slope
from the partial mandible. Further investigation uncovered the remaining
skeleton, lying supine in an extended position with the feet still buried within
intact sediments in the north road bank (figs. 3a and 3b, see color section following page 48).

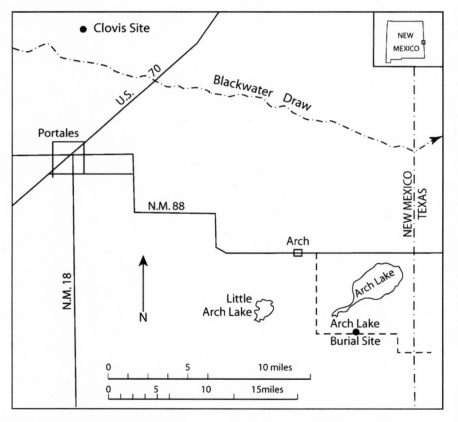

FIGURE 1. Location map of the Arch Lake burial in eastern New Mexico showing the
Arch Lake basin and the Clovis type site in Blackwater Draw.

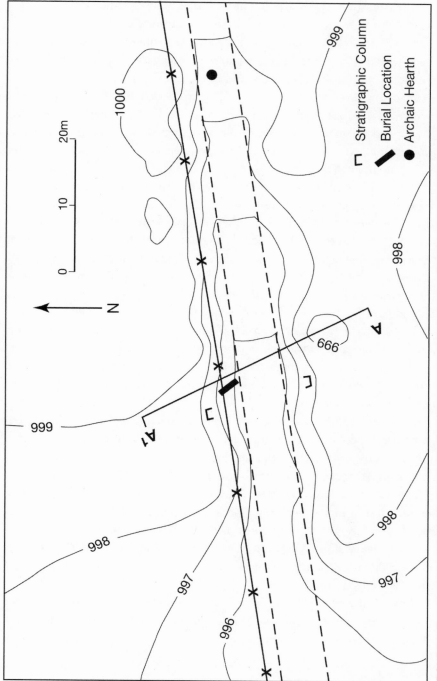

FIGURE 2. Topographic map of the Arch Lake site showing location of the burial pit, geological cross section, Archaic hearth, and stratigraphic columns.

The exposed skeleton was left in situ. That evening James Warnica, president of the El Llano Archaeological Society of Portales, was contacted, and the three men returned the following day. The geologic context of the interment, with its well-made flake tool and red staining, convinced Warnica of the probable antiquity of the remains, and he requested assistance from Dr. F. Earl Green of Texas Technological College. Due to concern that the skeleton would be damaged or destroyed by vandalism or subsequent weathering, Green recommended its removal in a single block of matrix. Notes on the burial and geologic context were recorded and the position of the skeleton was sketched and photographed.

The longitudinal axis of the burial was oriented N38°W, with the head to the southeast and feet to the northwest. The sidewalls of the purposefully dug grave were nearly vertical, but whether the ends were vertical or tapered was not observed. Except for the northwest end, most of the sediments overlying the interment were eroded, making it impossible to record the precise length of the burial pit. The depth and width of the pit were clearly discernible from the exposed geologic section of the road cut (figs. 3a and 3b, color plates). The pit width varied from 27.6 cm to 35.5 cm; the depth was uniform at 1.1 m. Green's stratigraphic profile shows that the burial trench was excavated from an ancient surface that was subsequently buried by 3 to 5 feet (0.9 to 1.5 m) of brown eolian sand. The contact is shown as being sharp, but there is no indication as to whether any erosion of the burial pit had occurred before being buried by later dune sand.

A strong red hematite stain ran along the outside and below the left humerus (fig. 4, color plate). A unifacial stone tool was found within this ocher concentration near the junction of the left humerus and radius. This placement suggests that the tool and red pigment were contained within a perishable pouch at the woman's waist. Fourteen disc-shaped beads were spaced in an arc about 6 inches (15 cm) wide located just above the clavicles (Gregg Moore, pers. comm. . 2000). Screening the burial's back dirt resulted in the recovery of five additional beads. A large cylindrical, nonhuman bone was found above right ribs 8, 9, and 10 (fig. 4). This was probably a tool but is missing from the collection.

In 1967 the human bones were coated with shellac and a wooden frame was built around the dirt pedestal. Gaps were filled with plaster and the human remains were covered with cotton batting. Then the matrix block and wooden frame were wrapped with Plaster of Paris–soaked burlap. After careful preparation for transport, the burial was taken to Eastern New Mexico University for curation.

Conservation

From February 17 to 20, 2000, the block containing the in situ skeletal remains was examined at the Blackwater Draw Museum in a laboratory environment that enabled cleaning, direct observation, sediment sampling, and controlled excavation of the cranium, lower left ribs, pelvis, and limb bones.

Prior to this study the skeleton and its surrounding matrix had been stabilized with a variety of solvents beginning with white shellac applied in the field in 1967. In 1994 the dirt surrounding the consolidated skeleton was saturated with Butvar B-98. B-76 with acetone was used as glue in matrix cracks that developed in 1995. Still, pieces of the matrix were gradually falling away from the consolidated portions of the skeleton.

To permit study and measurement of the human remains, it was first necessary to remove a tenacious and obstructing coating of sediment, calcium carbonate, and preservatives. Various solvent applications were used to soften this adhesive veneer surrounding the bone. Smithsonian conservator Carolyn Rose determined that ethanol successfully dissolved the shellac. However, more of the skeleton had been impregnated with the B-98 than anticipated. Acetone and denatured ethyl alcohol also were applied to dissolve consolidants from bone surfaces. A poultice of acetone-soaked laboratory tissues was used to soften the consolidated matrix block to level it out for photography. For heavily cemented areas, a 40% toluene solution was also used.

STUDY OF
THE SITE

Human Osteology

✦

THE burial is an extended position interment: the skeleton is supine, legs fully extended, face oriented skyward and tipped slightly to the right or east (figs. 5 and 6, color section). The head lies to the southeast. The partially mineralized skeleton is slightly elevated on the left. The cranium is in poor, fragmentary condition, as are the ribs and the pelvis. The long bone diaphyses of the humeri, femora, tibiae, and fibulae are well preserved.

Dental development and distinguishing characteristics of the bone identify this individual as a female aged seventeen to nineteen years. The proximal epiphyses of the tibiae appear to be united, the left ischial tuberosity is visible and unfused, and the posterior iliac crests are visible and not attached. The mandible is gracile with a wide gonial angle and no gonial eversion. There are no supramastoid ridges and the mastoid processes are small. There is no development of the supraorbital ridges and the superior orbital rims are sharp. Muscle attachment areas such as the deltoid tuberosities on the humeri and lineae aspera on the posterior femora are only slightly developed.

The cranium is missing the left anterior frontal, left malar, zygomatic arch, and anterior temporal squamous. The U-shaped maxillary dental arcade is in place, but the facial surface alveolar bone has eroded away. The left posterior body and ascending ramus of the mandible show rodent gnaw marks on the left condyle and in the area between the condyle and coronoid process. This portion of the mandible was originally found in another location near the burial, probably displaced by rodents, and had been reattached but out of normal anatomical position. It was removed for examination. The symphysis and the right half of the mandible are missing.

The cranium is large with a relatively short upper facial height and relatively narrow nasal aperture and slight alveolar prognathism (figs. 7a–e). The right zygomaxillary suture is evident and does not show recurvature. Cranial vault sutures are open but were difficult to see before cleaning due to adhering matrix.

9

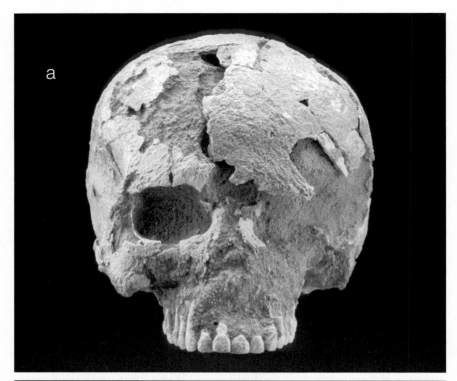

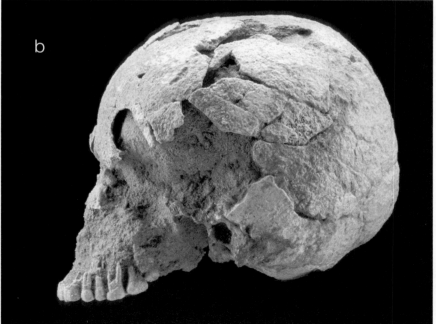

FIGURE 7. Frontal (a), left lateral (b), right oblique (c), right lateral (d), and posterior (e) views of the cranium positioned in the Frankfort Horizontal.

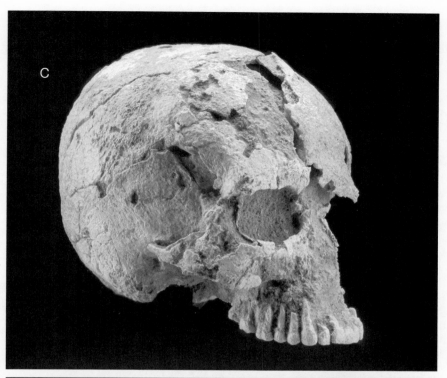

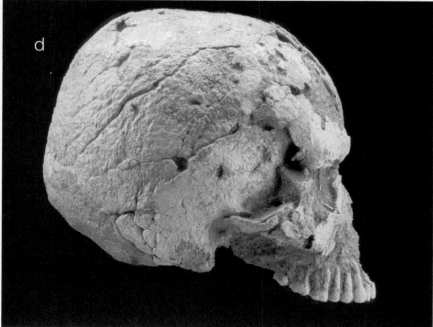

FIGURE 7 (continued), right oblique (c), and right lateral view (d).

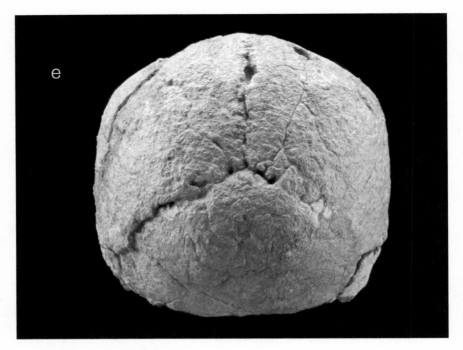

FIGURE 7 *(continued)* posterior view (e).

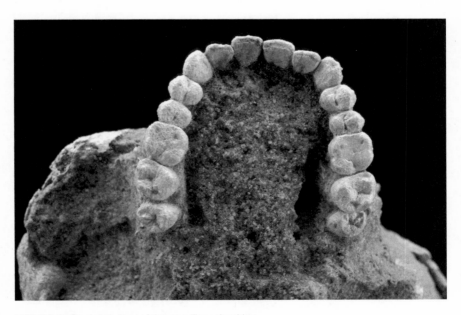

FIGURE 8. Occlusal view of the maxillary dentition.

The maxillary dentition is present in normal anatomical alignment without crowding (fig. 8). The teeth are held in position by the sandy matrix; the sockets, with supporting alveolar bone, and the remainder of the maxillae have eroded away. The crowns are well preserved; however, not all of the roots are present, especially those of the anterior teeth. The right third molar is fully erupted and level with the occlusal plane; the left is unerupted. The maxillary incisors show only trace shoveling and no double shoveling. No cusps of Carabelli are present. Wear is moderate on the incisors and slight on the first molars. Seven mandibular teeth are present, including three loose teeth (left canine, left first premolar, and right second premolar) that were collected in the field and integrated into the analysis. On the left are the canine, the crown and partial root of the first premolar, and all three molars. The molars are in the alveolar sockets in the same occlusal plane (fig. 9). Third molar development was at least root complete (R_c) in its formation (Moorrees et al. 1963). On the right are the crowns and partial roots of the second premolar and first molar. The tooth enamel is etched and finely irregular from postmortem leaching of calcium, and remnant amounts of calcium carbonate adhere to the teeth.

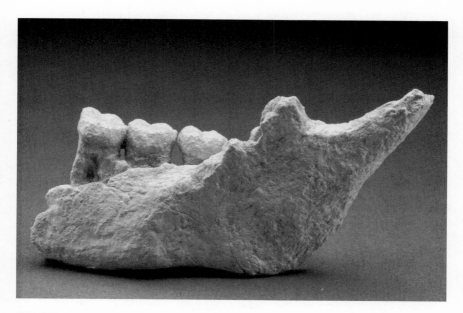

FIGURE 9. Left posterior body of the mandible after removal of a heavy carbonate encrustation.

The cervical vertebrae are in proper anatomical position, with C1 through C3 on an incline leading up to the base of the cranium. Both humeri are extended at the sides. There is little preservation of the right clavicle and the joint surfaces of the humeri. The left humerus, which is positioned higher in the matrix than the right, has rotated so that the anterior surface is down. The left forearm, represented by the proximal third of the diaphysis of the radius, indicates that the left arm was flexed at the elbow, which would have positioned the missing hand across the abdomen.

Red pigment is concentrated in an area lateral to the distal half of the left humerus, inferior to the flexed elbow and along the distal left femur (fig. 10, color section). The lateral posterior surface of the proximal third of the left radius is discolored red; this is not as evident on the medial surface. The posterior distal half of the left humerus is similarly stained red. The pigment extends parallel and lateral to the humerus to approximately mid-diaphysis. This stain is also present in the cross section through the lumbar region, where it is much more diffuse. No other bones appear to be stained.

The right upper arm is at the side but with slight deviation away from the body (slight lateral deviation distally); the lower right arm is not observed. The forearms are not well preserved, although a few finger bones were collected at the time of discovery.

Both legs are extended. The left femur, tibia, and fibula are in normal anatomical positions. The right femur has rotated laterally, as have the right tibia and right foot. The left foot has deviated medially. Both femora and tibiae are well developed and there are no obvious abnormalities. The femora have large lesser trochanters and no platymeria. The long bones show no evidence of cortical thinning. The five metatarsals on the left side are in place, but with medial deviation and the plantar surface down. An outline remains of the phalanges of the left foot. The right tarsals are poorly preserved; three metatarsals (3, 4, and 5) are present in a hyper-extended position (plantar flexion) with the lateral surface down.

The position of the skeleton with the left side slightly elevated, the head tilted toward the right, and more relaxed lateral displacement of the right arm indicate that the body was off center within the burial shaft and against the left wall.

Geoarchaeology

✦

THE exact location of the burial in the north wall of the road cut has been obliterated due to attrition during the past thirty-three years (fig. 2). However, the stratigraphy in the north and south walls remains intact. Stratigraphic profiles were exposed and described on both walls of the road cut (figs. 11a, 11b, and 12), and sediment samples were collected (fig. 13, color plate). Additional geological testing at the site was undertaken to determine whether additional skeletal remains were present. The results were negative, as described later.

The south wall reveals a column of three reddish brown dune sand strata (figs. 11a, 12, and 13; strata Z2, Z1, and Y) that overlie a light brown to white eolian sand (stratum X) into which the burial pit had been excavated (fig. 11b). Both Z1 and Z2 are weakly consolidated fine- to medium-grained eolian sands bearing uneven, irregular light and dark laminations and weakly developed pedogenesis (table 1). Stratum Z1 is absent from the north side column.

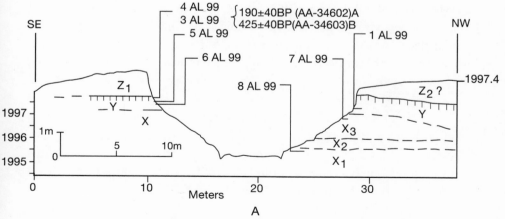

FIGURE 11A. Geologic cross sections through road cuts at the Arch Lake burial site.

15

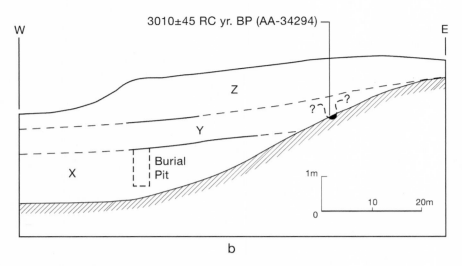

FIGURE 11B. Generalized profile of the north bank of the road cut.

Except for transition Y/X, the basal contacts are sharp with little disruption by bioturbation. Stratum Y is a darker reddish brown eolian sand with a grayish brown desert A horizon over a weakly prismatic B horizon. The basal contact is very irregular, with the top of stratum X having numerous filled rodent burrows (bioturbation). Stratum Y was sampled from the south wall section (figs. 12a and 13; samples 3, 4, 5, and 6). Radiocarbon measurements on sample 3AL99 from stratum Y were 190 ± 40 RC yr. (AA-34602) on bulk sediment and 425 ± 40 RC yr. (AA-34603) on humic acids. These dates provide a minimum age of 400 RC yr. for pedogenesis following the deposition of stratum Y. It is apparent that nuclear-age carbon has contaminated this paleosol, making it appear anomalously young for the degree of pedogenesis. A weaker soil occurs at the top of stratum Z1.

On the north wall of the road cut, the oldest dune deposit, Stratum X, is subdivided into three parts (fig. 12b). The lowest part, stratum X1, is laminated medium- to coarse-grained light brown sand without visible $CaCO_3$. This is transitional upward over 3 cm to stratum X2, a very light brown (white), firm, calcareous, laminated medium- to coarse-grained sand with thin (1–2 mm) $CaCO_3$ laminae and mottles. Stratum X2 is gradational upward over 5 cm to stratum X3, which is composed of light brownish white firm, massive medium- to coarse-grained sand with lighter and darker mottles due to bioturbation. It is clear from Green's 1967 stratigraphic sketch and photographs (figs. 3a and 3b) that the burial pit was excavated from the Y/X contact and penetrated strata X3 and X2, ending in the top of X1.

Several small lumps of dispersed charcoal (sample 1AL99) collected from the heavily bioturbated irregular contact between strata X and Y could be from a buried hearth, but the intense bioturbation by rodents makes the stratigraphic provenance questionable. Thus dating this charcoal was not attempted. An in situ radiocarbon sample (2AL99) was collected from a distinct hearth feature exposed in the road bed about 40 m east of the burial site (fig. 2). The hearth pit was reportedly thicker in the past, but the top of the fire pit has been lost through road grading and deflation. The surface from which the fire pit was excavated cannot be determined with certainty (fig. 11b). The hearth appears to have been within stratum Y because artifacts have been eroding out of this stratum. An AMS radiocarbon date of 3010 ± 45 RC yr. BP (AA-34294)

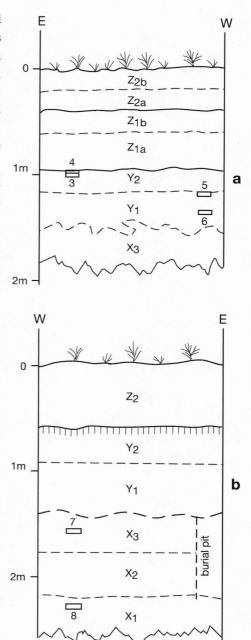

FIGURE 12. Stratigraphic columns for the south and north banks of the road.

indicates a Late Archaic age for the hearth. Whether stratum Y deposition was complete before the fire pit was dug is unknown. However, it does provide a minimum age for the beginning of Y deposition.

On the basis of color, calcification, and stratigraphic as well as geomorphic

TABLE 1. SEDIMENTARY CHARACTERISTICS OF STRATA EXPOSED BY THE ROAD CUT AT THE ARCH LAKE BURIAL LOCALITY. DESCRIPTIONS BY HAYNES.

Stratum		Description	Thickness (cm)
Z_{2b}	Sand	Light brown (10 YR $4/_3$) soft, massive, fine to medium eolian sand with weak medium blocky pedogenic structure. Basal content gradational over ~ 3 cm.	20
Z_{2a}	Sand	Light brown (10 YR $4/_3$) soft, laminated, fine to medium eolian sand with weak, very coarse, pedogenic structure. Sharp (erosional?) basal contact.	60
Z_{1b}	Sand	Light brown (10 YR $5/_4$) soft, massive, medium to coarse eolian sand. Sharp, irregular basal contact.	20
Z_{1a}	Sand	Light brown (10 YR $5/_4$) soft banded fine to medium (darker) and medium to coarse (lighter) sand. Sharp (erosional?) basal contact.	35
Y_2	Sand	Grayish brown (10 YR $5/_{3.5}$) firm, friable, fine to medium silty sand. Upper 2 cm. is grayer (desert A-horizon?). Basal contact transitional over 2 cm. Samples 3 and 4 AL 99 from upper part (see figs. 12 and 13).	30
Y_1	Sand	Reddish brown (7.5 YR $4/_3$) firm, friable, massive, medium to coarse eolian sand with some clay. Very irregular, bioturbational basal contact zone 10–20 cm. thickness. Sample 5 AL 99 from top and sample 6 AL 99 from lower half (see figs. 12 and 13).	50
X_3	Sand	Mottled light brown, reddish brown, and white (10 YR $8/_3$), calcareous, firm, friable, massive medium to coarse sand. Basal contact gradational over ~ 5 cm. Sample 7 AL 99 from upper half (see fig. 12).	40
X_2	Sand	Very light brown (10 YR $7/_2$) to white, firm, friable to soft, laminated, medium to coarse sand with $CaCO_3$ bands and blotches. Bioturbated in places. Basal contact gradational over ~ 3 cm.	40
X_1	Sand	Light brown (10 YR $7/_2$), soft, laminated, medium to coarse sand without $CaCO_3$ bands, otherwise the same as X_2. Basal contact not observed. Sample 8 AL 99 from near top (see fig. 12).	>50

position, stratum X is assumed to be no younger than late Pleistocene, 11,500 RC yr. BP (ca. 13,500 CAL yr. BP). The stratum Y dune sand and paleosol appear to be late middle Holocene on the basis of their stratigraphic position and pedogenic development. The Altithermal climatic episode (Antevs 1955) is likely represented by the erosional hiatus of the Y/X contact. The paleosol developed in stratum Y very likely represents more mesic conditions that followed the Altithermal. The stratum Y dune-stabilizing soil is a paleosol that has a widespread occurrence in the High Plains (Scott 1963; Malde and Schick 1964; Haynes 1968; Holliday 1995; Muhs et al. 1996).

At the Clovis type site in Blackwater Draw, the Altithermal erosion occurred between 8500 and 6500 RC yr. BP (Haynes 1995; ca. 9500 to 7300 CAL yr. BP). The stratum G1 dune sand at Blackwater Draw is a possible correlate of stratum Y at the Arch Lake burial site. Stratum G1 also correlates with Holliday's (1995) stratum 4s, which contains his Lubbock Lake paleosol that began to stabilize the 4s dunes by 4500 RC yr. BP. (ca. 5200 to 5100 CAL yr. BP) The Y/X erosional surface probably formed by deflation during the Altithermal period. If these correlations are correct, the burial at Arch Lake has a minimum age of 8000 RC yr. (ca. 9000 to 8800 CAL yr. BP). If stratum X dune sand accumulated during the Clovis drought that terminated the Pleistocene epoch (Haynes 1991), the burial could be as old as 11,000 RC yr. (ca. 13,000 CAL yr. BP).

Subsurface reconnaissance via bucket auger at the Arch Lake site explored the possibility of additional burials existing north of the primary discovery location. The site geology consists of approximately one meter of fine to medium eolian sand that disconformably overlies a calcic horizon and soil C horizon (stratum X) where the human remains were recovered. The calcic zone (stratum X_2 of Haynes) is a Bcab2 horizon that is a prominent physical horizon clearly appearing in 1967 excavation photographs. The skeleton's depth below the modern ground surface and into the calcic horizon was reconstructed using these photographs.

The top of the skeleton was 1.1 m (42 inches) below the contact of the upper soil C horizon (stratum Y_1) and the top of the calcic (stratum X_2) horizon. Therefore the top of the skeleton would have been approximately 2.4 m below present ground surface. Based on these depth reconstructions, auger test holes were drilled in order to examine the sands for traces of red staining similar to the hematite (red ocher) present on some bones and sediments surrounding the human burial. The unaltered matrix sands were light brownish gray (10YR6/2d) and contrasted significantly with the ocher-stained burial sands, which were pink (7.5YR7/4d) to light reddish brown (10YR6/4d).

Nineteen hand-augered holes, each 10 cm in diameter and approximately 3 m deep, were dug in an area that extended 10 m north, 5 m west, and 14 m east of the Arch Lake burial's discovery point. An additional eight holes were augered up to 50 m north from the site and at 5 m apart in a north-south line.

No ocher-stained sands were encountered in any auger hole. The north-trending series of holes revealed that the calcic horizon (stratum X_2) decreased in thickness northward. Geologically there is no present evidence for additional burials at the Arch Lake site. However, ground-penetrating radar, an approach that has not been used, would be well suited to the Arch Lake site because the target horizon is close to the ground level and the sands are porous, well sorted, and dry.

Microprobe Analysis of the
Iron-bearing Sediment

✦

THREE samples of burial fill taken in the area of dense red pigment between and below the distal left humerus and left ribs were studied using a petrographic microscope and a JEOL 8900 electron microprobe under the following conditions: (1) an acceleration voltage of 15 kv; (2) a beam current of 10 nA; and (3) a beam diameter of 10 μm. The purpose was to identify the composition of the reddish stained sands. The following standards were used: (1) USNM 143965 for Na, K, and MG; (2) USNM R-2912 for Si, Al, and Ca; (3) USNM 157872 for Mn; and (4) USNM 10421 for P. The results were normalized to the nominal compositions of the standards and adjusted to represent 100% of the sample.

Samples designated AL-11 and AL-20 consisted of unlithified and very poorly sorted quartz grains ranging in size from coarse sand (1–0.5 mm) to silt (63–30 μm). All of these samples have a chalky, reddish brown, ferruginous calcareous coating. Sample AL-12 differed only in that there was enough calcareous material to consolidate the quartz grains into a weakly cemented sandstone (fig. 14, color section).

The microprobe analyses of this coating on the grains and the matrix in sample A-12 (table 2) indicate high quantities of calcium, silica, and iron. These concentrations would be expected from calcareous ferruginous quartz sand. This iron-rich, calcareous, silty sandstone with its reddish brown coloration could be considered to be a type of sediment-rich ocher.

21

TABLE 2. ELECTRON MICROPROBE ANALYSIS OF ARCH LAKE SOIL SAMPLES (WT.%).

Sample	n		SiO_2	Al_2O_3	FeO	MgO	MnO	CaO	K_2O	Na_2O	P_2O_5	Total
Sample AL-11	7	Mean	34.6	10.6	5.7	2.7	0.5	43.5	1.2	0.9	0.2	100
		s. d.	8.1	4.4	0.7	0.3	0.9	12.1	0.3	1.0	0.1	
Sample AL-20	18	Mean	20.2	6.4	4.1	2.8	0.1	64.7	0.8	0.6	0.4	100
		s. d.	12.1	3.9	2.1	1.4	0.2	20.5	0.4	1.9	0.3	
Sample AL-12	9	Mean	36.1	8.9	5.8	2.8	0.5	43.6	1.6	0.6	0.2	100
		s. d.	11.1	2.2	1.7	0.7	1.4	13.5	0.5	0.3	0.1	

Radiocarbon Dating of the Skeleton

✦

G EOCHRONOLOGY studies of the burial site included radiocarbon measurements and chemical analyses of the skeleton. The primary objective was to determine the geologic age of the skeleton by directly dating the human bone. Two samples were tested for potential dating: 4 g of bone removed from the midshaft of the right femur and a right mandibular first molar. The suitability of the femur or molar for dating was assessed through quantitative amino acid analyses, which determined the quantity and quality of protein preserved.

Table 3 summarizes the amino acid compositions of the human bone and tooth dentin compared to a modern sample. The protein (collagen) content of the femur is approximately one-tenth that of a modern bone; the tooth has one-tenth the protein of the ancient human femur, i.e., about 1% of modern. Modern bone and dentin contain approximately 2500 nanomoles (nmol) of total amino acids (AA) as collagen per milligram of bone. The femur contained 326 nmol AA's/mg or 13.0% of a modern bone. The tooth dentin contained 26 nmol AA's/mg or 1.1% of modern.

Although there have been significant losses of protein from the Arch Lake human remains, the chemical composition of the remaining protein is more suitable for dating than would be indicated from the initial "nmol AA/mg" values. The amino acid composition of the bone is "collagenous": the amino acid ratios are similar to collagen except for two amino acids, proline and glutamic acid, with proportions that are slightly altered relative to modern collagen.

The tooth dentin has a collagen-derived composition because there are moderate changes in the R/1000 values of five key amino acids: hydroxyproline, aspartic and glutamic acids, proline, and arginine. The tooth's amino acid composition is derived from collagen but not the same as collagen. Although the total amount of protein is significantly reduced, the amino acid spectrum of this protein is relatively well preserved despite oxidative diagenesis during burial.

Both the cortical bone and tooth dentin were suitable for accurate ^{14}C dating because their remnant protein resembled collagen. However, too few

TABLE 3. QUANTITATIVE AMINO ACID ANALYSES OF ARCH LAKE HUMAN BONE AND TOOTH SAMPLES.

AMINO ACID	MODERN BONE R/1000	SR-5259 FEMUR R/1000	SR-5260 TOOTH DENTIN R/1000
Hydroxyproline	93[a]	96	69
Aspartic Acid	50	55	93
Threonine	19	18	21
Serine	33	13	18
Glutamic Acid	79	75	105
Proline	115	145	96
Glycine	327	336	336
Alanine	113	115	125
Valine	20	27	37
Methionine	11	0	0
Isoleucine	14	11	17
Leucine	31	24	30
Tyrosine	6	0	3
Phenylalanine	14	12	14
Histidine	8	0	3
Hydroxylysine	8	3	8
Lysine	28	30	23
Arginine	31	40	4
Total nmols of amino acids per mg sample	2500	326	26
% of modern bone or tooth	100	13.0	1.1
Amino Acid Composition	Collagen	Collagenous	Collagen-derived

[a] Amino acid proportions are expressed in residues per thousand (R/1000) and as nanomoles of each amino acid per milligram (nmol/mg) of modern bone or dentin.

grams of dentin were available, and the tooth would not have yielded enough purified amino acid carbon for AMS dating. Therefore only the cortical bone was processed further. The techniques for chemical purification and dating of bone and the multiple-fraction dating protocol are discussed in Stafford et al. (1991), Stafford (1998), and Stafford et al. (1999).

Specimens such as the Arch Lake skeleton require more stringent proof of their geologic age because ancient human skeletons are extremely rare and because the collagen content was low enough that one ^{14}C date would not have been definitive evidence for geologic age. Dating several different chemical fractions that are progressively purer chemically enables the removal of foreign carbon contamination to be monitored. If a sample's ^{14}C age increases significantly with each progressive purification step, the presence of significant amounts of markedly different exogenous compounds is indicated. If the ^{14}C age changes by only a few hundred years with progressive purification, or if several different chemical fractions yield the same age, less contamination is indicated, and there is increasing certainty that the final age on the chemically purest fraction is the accurate age.

The cortical bone was removed by hand-sawing using a brass jeweler's saw blade 0.7 mm wide. The exterior of the bone was brown with adhering sediment. The interior cortical bone was white, hard, very slightly chalky, and had a matte surface. The medullary cavity was filled completely with silty, very fine-to-fine quartz sand. The bone sample removed measured 20 mm long by 22 mm wide by 11 mm thick and weighed 4 grams. After removing the outer 1 mm on all surfaces, 1.8 grams were used for chemical purification and radiocarbon dating.

After decalcification in 4°C 0.2N HCl and extraction in 4°C 0.1% KOH, 58.1 mg of KOH-collagen were obtained. This 3.3% yield compares to a 20% yield for modern bone. A fraction of the KOH-collagen was heated at 110°C in pH 2 DI water to isolate the "gelatin" fraction. The filtered, hot water–soluble collagen was finally hydrolyzed in 110°C 6N HCl and purified by passing the hydrolyzate through a column of 0.2 mm diameter XAD-2 resin. The hydrolyzate was dried over N_2 gas before combustion for ^{14}C dating.

The goal was to date several different chemical fractions from the least to the most chemically pure. Four AMS ^{14}C age measurements were made on the Arch Lake femur. The measurements were: (1) the first fraction obtained, the KOH-collagen; (2) gelatin from the KOH-collagen; (3) the XAD-purified hydrolyzate of the KOH-collagen; and (4) the XAD-purified hydrolyzate of gelatin isolated from KOH-purified collagen. Finally, a stable isotope measurement was made on the gelatin fraction. These measurements are summarized in table 4.

For the Arch Lake samples, the ^{14}C ages increase as chemical purification improves. The KOH-collagen and the purer gelatin yielded ages of 9295 ± 35 RC yr. (CAMS-80539) and 9700 ± 40 RC yr. (CAMS-69916), respectively. Hydrolysis and XAD purification removed more contamination, predominantly from fulvic acids. Consequently both XAD dates were older than preceding fractions. The XAD-KOH-collagen measured as 9885 ± 45 RC yr. (CAMS-80540), and the theoretically purer of the two XAD fractions, the XAD-gelatin-KOH-collagen, dated 10,020 ± 50 RC yr. BP (CAMS-61133). Although the two XAD fractions overlap at two standard deviations, the 10,020 ± 50 RC yr. age is considered the more accurate and best estimate of the Arch Lake human burial's geologic age, which has a 2-sigma calibrated range of 11,950 to 11,200 CAL yr. BP.

Oxford University previously measured ^{14}C ages for the same femur and a molar; the results are summarized in table 4. The Oxford measurements range from 8500 ± 100 RC yr. (OxA-3824) to 9160 ± 100 RC yr. (OxA-3823). The average of these measurements and the date used as Oxford's age estimate for the skeleton is 8870 ± 40 RC yr. BP.

This study's estimate of the geologic age of the Arch Lake human skeleton (10,020 ± 50 RC yr., CAMS-61133) is used as the skeleton's geologic age because the XAD-purified hydrolyzed gelatin from KOH-extracted collagen has theoretically the least probability of containing exogenous organic matter. Although the 9885 ± 45 RC yr. age on the first XAD fraction overlaps the second date at two standard deviations, the gelatin extraction step would have removed contaminants remaining after the KOH purification alone. The similarity of ^{14}C ages on less chemically pure KOH-collagen and gelatin is an indication that exogenous geologically younger carbon is not abundant. The XAD purification step apparently removed traces of fulvic acids, which were not eliminated until the hydrolyzed collagen was passed through XAD resin.

The 1150 RC year difference between the two laboratories' results (8870 ± 40 RC yr. at Oxford versus 10,020 ± 50 RC yr. in this study) is probably due to the different methods for removing trace amounts of geologically younger, small molecular weight organic carbon contamination. The retention of fulvic acids bound to collagen and collagen-derived peptides is the most probable source of foreign contamination.

TABLE 4. AMS RADIOCARBON MEASUREMENTS ON ARCH LAKE FEMORAL BONE AND TOOTH DENTIN.

Sample Description	Chemical Fraction Dated	14C AGE RC yr. ± 1SD	δ13C/‰ (PDB)	δ15N/‰ (AIR)	Lab No.	1 sigma Calibrated Date CAL yr. BP	2 sigma Calibrated Date CAL yr. BP
			OXFORD UNIVERSITY AMS 14C DATES				
Right Femur	Ion-exchange purified gelatin hydrolyzate	8680 ± 160	-16.4‰	---	OxA-3265[a]	10,150-9500[b]	10,200-9400
Right Femur	Chloroform-extracted gelatin	9090 ± 100	-15.1‰	+12.6	OxA-3822	10,470-10,150	10,550-9900
Right Femur	Chloroform-soluble residue ("Shellac")	9160 ± 100	-15.1‰	---	OxA-3823	10,480-10,220	10,700-10,150
Molar	> 10kD gelatin	8500 ± 100	-17.6‰	---	OxA-3824	9600-9320	9750-9250
AVERAGE		8870 ± 40				10,160-9890	10,180-9770
			STAFFORD RESEARCH LABORATORY AMS 14C DATES				
Right Femur	KOH-Collagen	9295 ± 35	-17.96	+8.4	CAMS-80539[c]	10,560-10,420	10,640-10,280
Right Femur	XAD-KOH Collagen	9885 ± 45	---	---	CAMS-80540	11,340-11,200	11,550-11,180
Right Femur	Gelatin	9700 ± 40	-13.2	+13.5	CAMS-69916	11,190-10,910	11,210-10,790
Right Femur*	XAD Gelatin (KOH-Collagen)*	10,020 ± 50*	---	---	CAMS-61133*	11,640-11,260*	11,950-11,200*

* Indicates Chemical Fraction believed to represent the most accurate geologic age estimate

[a] OxA = Oxford Accelerator, Research Laboratory for Archaeology and the History of Art, University of Oxford, England.
[b] One sigma (68.2% confidence interval) and two sigma (95.4% confidence interval).
[c] CAMS = Center for Accelerator Mass Spectrometry, Lawrence Livermore National Laboratory, California.

Stable Isotope Values

✦

S TABLE carbon and nitrogen isotope measurements provide data regarding the woman's diet and paleoecology. Interpretations have varying degrees of accuracy because (1) bone collagen is poorly preserved chemically and diagenetic effects alter original isotopic compositions, (2) direct isotopic data from fauna consumed by the human are missing, and (3) the region's past ecological changes between 8000 and 12,000 RC yr. ago will have shifted the base isotopic values for floral communities over time. Consequently, it is difficult to distinguish between trophic level and paleoecological-derived isotopic changes.

Isotopic values used for paleodiet interpretation differ between the two labs and among chemical fractions analyzed (table 4). All Oxford $\delta^{13}C$ values range from −15.1‰ to −17.6‰; Oxford's $\delta^{15}N$ value was 12.6‰. Stafford's $\delta^{13}C$ values range from −13.2‰ to −18‰; $\delta^{15}N$ values varied from +8.0‰ to +13.5‰ (table 4).

The two labs used different chemical fractions as the one most accurate for radiocarbon dating. The respective C and N isotope values on Oxford's selected fraction, "chloroform-extracted gelatin," were −15.1‰ and +12.6‰. The analogous values from Stafford's analyses were −13.2‰ and +13.5‰ for C and N isotope values, respectively. Both labs' values were from the right femur, and averaging these measurements yields a $\delta^{13}C$ value of −14.2‰ and a $\delta^{15}N$ value of +13.1‰ for the Arch Lake woman.

A collagen $\delta^{15}N$ value of +13.1‰ reflects a diet nominally with a high proportion of animal protein. This value is that of a high trophic level animal (carnivore) that consumed a significant proportion of animal protein relative to plant protein. Due to the continent-wide extinction of the late Pleistocene megafauna, the primary large herbivore available for hunting on the southern High Plains after 10,800 RC yr. BP would have been the American bison, Bison sp. Because modern, temperate climate, terrestrial herbivores have $\delta^{15}N$ values ranging from +1.8 to +6.1‰ (Bocherens 2000) and each trophic level

28

represents an approximate 4‰ increase in δ^{15}N (Bocherens 2000), the Arch Lake human's δ^{15}N value would represent at least two trophic levels above the herbivorous bison. However, assessing trophic level (degree of carnivory) requires that the absolute carbon and nitrogen isotopic values be known for the animals hunted. These data are not common and are sometimes equivocal. Stafford has unpublished isotopic data on early Holocene (9000–11,000 RC yr. BP) bison from Lubbock, Texas, 145 km southeast of Arch Lake. The δ^{15}N values on Texas bison bone collagen vary from +5 to +11‰. This range in values from the fossil record complicates finding a single δ^{15}N value to use for bison if it was the primary food source. Compounding the difficulty of achieving precise estimates for human diet is the effect of diagenesis on the collagen's isotopic composition. Because δ^{13}C and δ^{15}N become > 2‰ more negative (isotopically lighter) due to diagenesis (Dobberstein et al. 2009), the isotopic values measured individually by Oxford and Stafford may be less accurate than desired due to the human bone's low collagen content.

Based on present data, general isotopic reconstructions for the Arch Lake woman's diet and paleoecology are (1) her diet had a high proportion of meat based on δ^{15}N values, and (2) the δ^{13}C value of bone collagen is consistent with a region having C_3/C_4 plant species compositions similar to that of the early Holocene in the Texas Panhandle.

Cranial Morphometrics

O NLY a limited cranial morphometric analysis was possible because of distortion during burial, as described elsewhere in this report. Measurements obtainable from Arch Lake are presented in table 5. Many of these are approximations, but ones in which there is considerable confidence. It is immediately evident from the values in table 5 that the Arch

TABLE 5. CRANIAL MEASUREMENTS FOR ARCH LAKE.

Measurement	Abbr.	mm
Glabella-occipital length	Gol*	(171)
Basion-bregma height	Bbh	(136)
Maximum cranial breadth	Xcb*	146
Biauricular breadth	Aub*	120
Biasterionic breadth	Asb*	116
Nasion-prosthion height	Nph*	64
Alveolar breadth	Mab	67
Alveolar length	Mal	58
Orbit height	Obh*	(31)
Orbit breadth	Obb*	(40)
Interorbital breadth	Dkb*	20
Inferior malar length	Iml	36
Maximum malar length	Xml	54
Cheek height	Wmh*	26
Frontal chord	Frc*	(102)
Frontal subtense	Frs*	(20)
Frontal fraction	Frf	(59)
Parietal chord	Pac*	113
Parietal subtense	Pas*	30
Parietal fraction	Paf	59
Occipital chord	Occ	97
Occipital subtense	Ocs	26
Occipital fraction	Ocf	44
Nasion radius	Nar	(80)
Bregma radius	Brr	(118)
Lambda radius	Lar	111
Opisthion radius	Osr	45

*Measurements used to compare Arch Lake to seven other Paleoamerican crania: Spirit Cave, Wizards Beach, Gordon Creek, Horn Shelter No. 2 (male), Browns Valley, Pelican Rapids, and Buhl.

Lake cranium departs from other ancient American crania in that it is short and wide. The cranial index is 85.3, placing it in the brachycranic range. Thirteen of the measurements presented in table 5 (those marked with *) were used to compare the Arch Lake cranium to seven other Paleoamerican crania (Spirit Cave, Wizards Beach, Gordon Creek, Horn Shelter No. 2, Browns Valley, Pelican Rapids [Minnesota Woman], and Buhl). Buhl measurements were obtained from Green et al. (1998) and Herrmann et al. (2006); all others were measured in the Howells system as described in Jantz and Owsley (2001). Measurements included were those common to the comparative sample. Analysis was performed as described in Jantz and Owsley (2001), using twenty-four world samples from Howells and two additional Native American samples (Blackfoot and Zuni).

Figure 15 shows the samples displayed on the first three principal coordinates, accounting for 63.17% of sample variation. Table 6 gives the structure coefficients indicating the variables that contribute to the principal coordinate scores. Arch Lake has an extreme positive location on the first axis, which primarily reflects a short, wide vault. In this feature it is set off from most other Paleoamericans, which have longer, narrower crania. On PC 2, Arch Lake has a strong positive score, along with several other Paleoamerican specimens, including Pelican Rapids, Buhl, Gordon Creek, and the adult male from Horn Shelter No. 2. This axis reflects vault breadth, particularly at the base, without regard to length. Arch Lake and the other Paleoamerican crania are similar to several recent Native Americans on this axis. Arch Lake has a low score on PC 3, which it shares with all other Paleoamerican specimens except Wizards Beach. This axis reflects face height and parietal length. Modern populations with low scores on this axis are Africans and Southwest Pacific populations, as opposed to Native Americans, who have high scores.

Table 7 presents Mahalanobis distances (D) of the early American fossils from each other, using the pooled covariance matrix from the modern groups as described in Jantz and Owsley (2001). These distances may be compared to a random expectation of the distance between two crania drawn at random from a single population with the specified covariance matrix (Defrise-Gussenhoven 1967). Arch Lake exceeds the random expectation for all crania except Gordon Creek, although the Horn Shelter No. 2 to Arch Lake distance is only marginally greater than the random expectation.

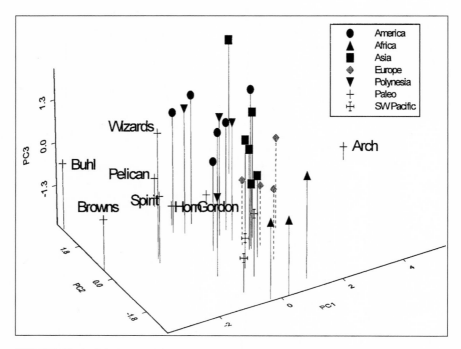

FIGURE 15. Arch Lake cranial measurements compared to modern world population samples and seven other Paleoamerican specimens.

Early American crania differ from modern Native Americans in having large vaults (Jantz and Owsley 2005). Vault size can be evaluated by using the geometric mean of cranial length, breadth, and basion-bregma height. Table 8 presents summary statistics for vault sizes of Arch Lake, three other female Paleoamerican crania (Gordon Creek, Buhl, and Pelican Rapids), and five modern Native American female samples. Arch Lake's size falls completely outside the range of variation of all modern groups except for the Blackfeet, where Arch Lake is near the maximum for this group. Comparing Arch Lake to the other Paleoamericans reveals astonishing homogeneity. Arch Lake, Buhl, and Gordon Creek are virtually identical. Only the Pelican Rapids individual, whose age is fourteen to fifteen years, is somewhat smaller, but still above the mean of all groups except the Blackfeet.

Morphometrically the Arch Lake cranium is unique among the early American crania in the sample for the present study in having a short, broad

TABLE 6. STRUCTURE COEFFICIENTS FOR THE FIRST
THREE PRINCIPAL COORDINATES

Variable	PC1	PC2	PC3
Gol	-0.6346	0.0294	-0.4617
Xcb	0.4695	0.7501	0.0319
Aub	-0.2806	0.8863	0.0547
Asb	0.4253	0.6578	-0.5479
Nph	-0.0625	0.4036	0.6535
Obh	-0.2027	0.1940	0.7302
Obb	0.0709	-0.0947	0.1248
Dkb	-0.2171	-0.4647	0.0856
Wmh	0.3244	0.6627	0.2025
Frc	-0.1982	-0.0418	0.0905
Frs	0.2720	-0.5968	0.0530
Pac	0.0557	-0.2744	-0.7104
Pas	0.7096	-0.3012	-0.1878

TABLE 7. MAHALANOBIS DISTANCES (D) BETWEEN EIGHT PAIRS OF
PALEOAMERICAN CRANIA.

	Pelican	Browns	Spirit	Buhl	Gordon	Horn #2 (Male)	Arch	Wizards
Pelican	0							
Browns	5.593	0						
Spirit	4.160	4.751	0					
Buhl	4.406	5.524	4.527	0				
Gordon	2.479	6.775	4.558	5.423	0			
Horn #2	2.794	4.947	3.672	5.054	3.018	0		
Arch	7.223*	9.694*	8.529*	10.031*	6.56	7.173*	0	
Wizards	5.031	5.075	3.169	4.642	5.928	4.83	8.640*	0

Note: Expected distance (D) = 5. Expected distance (D) of skulls drawn at random from a population with
the same covariance matrix as the pooled within matrix of reference samples. Distances greater than 6.96
can be considered significant.

TABLE 8. VAULT SIZE CALCULATED AS THE GEOMETRIC MEAN OF CRANIAL LENGTH, BREADTH, AND BASION-BREGMA HEIGHT OF FOUR PALEOAMERICAN FEMALES AND FIVE RECENT NATIVE AMERICAN SAMPLES

Group	N	Mean	s. d.	Minimum	Maximum
Arikara	27	143.55	3.27	136.89	148.44
Blackfeet	45	146.29	2.96	140.05	152.03
Peru	55	141.72	3.47	135.08	149.22
Santa Cruz	51	142.27	3.06	134.58	149.15
Zuni	23	139.74	3.47	132.70	145.35
Arch Lake	1	150.30	---	---	---
Buhl	1	150.32	---	---	---
Gordon Creek	1	150.30	---	---	---
Pelican Rapids	1	146.01	---	---	---

cranial vault. It is similar in this regard to many modern Native Americans and some Asians. Beyond that, it exhibits similarity to modern Native Americans in having a relatively broad base. Arch Lake differs markedly from modern Native Americans in having a low face and orbits, a feature it shares with other early American crania. It also possesses a large cranial vault, which sets it apart from modern Native Americans and aligns it with other early crania. Although Arch Lake is unique, it may be important that it is morphometrically most similar to Gordon Creek and Horn Shelter No. 2, which are closest to it geographically.

Dental Morphology and Measurements

✦

MEASUREMENTS and crown traits of the teeth were analyzed to gauge Arch Lake's similarity to recent Native Americans. Comparative data are from two arctic groups, nine groups representing major geographical regions in North America, and Peruvians from South America. Summary information for the subgroups of recent regional samples is presented in table 9. All specimens are curated by the Smithsonian Institution's National Museum of Natural History.

Utilizing the Arizona State University (ASU) dental anthropology system and referring to the focal traits and scaling system of Turner et al. (1991) and Scott and Turner (1997), fifteen discrete crown traits were examined for two accessible Paleoamerican crania with unworn teeth: Arch Lake and the juvenile from Horn Shelter No. 2. The presence or absence of the fifteen discrete traits for the two Paleoamerican tooth samples is given in table 10.

According to Turner (1984, 1987, 1990) and Scott and Turner (1997), North and South American Indians have the world's highest frequencies of winging (50%), shoveling (85–90%), and double shoveling (55–70%) in the maxillary central incisors. In the two Paleoamerican specimens, however, no winging or double shoveling is observed.

Interruption grooves on the upper second incisor along with odontomes on both upper and lower first and second premolars are most common in Native Americans (Scott and Turner 1997). These traits are not present in the two Paleoamericans. Yet traits that are present in the Paleoamerican specimens are not common in Native Americans. Specifically, the left upper first molar of Horn Shelter No. 2 has a tiny cusp of Carabelli, and both Paleoamericans have upper second molars with reduced hypocones. Appearance of the cusp of Carabelli on the upper first molar, reduction or absence of the hypocone on the upper second molar, and absence of the hypoconulid on the lower second molar are less common in Native Americans.

TABLE 9. GEOGRAPHIC ORIGIN OF DENTAL METRIC AND DISCRETE-TRAIT COMPARATIVE SAMPLES REPRESENTING ARCTIC POPULATIONS, NORTH AMERICAN INDIANS, AND PERUVIANS.

Sample name	Information
Aleuts	Mainly from Unalaska and Kagamil Islands, including Fox, Rat, and Near Islands
Subarctic	Canadian and Alaskan Indians
Alaska Eskimos	Lower Yukon and Kuskokwim River basins, Seward Peninsula, and Point Barrow
California	Alameda, Santa Barbara, and the Islands of Santa Cruz, San Miguel, and San Nicolas
Great Basin and Western Plains	Colorado, Idaho, Montana, Nevada, Utah, and Wyoming (male sample only)
Southwest	Arizona, New Mexico, and northern Mexico
Plains (Northern)	North Dakota, South Dakota, and Minnesota
Plains (Southern and Central Plains and Midsouth)	Arkansas, Iowa, Kansas, Louisiana, Missouri, Nebraska, Oklahoma, and Texas
Northeast Woodlands (East)	Connecticut, Delaware, Maryland, Massachusetts, New Jersey, New York, and Pennsylvania
Northeast Woodlands (Midwest)	Illinois, Indiana, Kentucky, Michigan, Ohio, and Wisconsin
Southeast Woodlands	Alabama, Florida, Georgia, Mississippi, North and South Carolina, Tennessee, Mississippi, Virginia, and West Virginia
Peruvians	Chicama, Chilca, Cinco Cerros, Nasca Region, San Damian, and other places

The other traits examined here are relatively infrequent in the New World region with the possible exception of the 6th cusp. The 6th cusp in the first lower molar is more common in Native Americans than it is in East/Southeast Asians. Neither the Arch Lake female nor the juvenile from Horn Shelter No. 2 has a 6th cusp.

To determine the degree of shoveling of the upper first incisors, which is one of the most important traits in the dental dichotomy of pan-Pacific populations, depths were directly measured. Lingual fossa depths for the two Pa-

TABLE 10. TOOTH CROWN TRAITS IN TWO PALEOAMERICAN DENTITIONS (BASED ON THE ASU SYSTEM).

	Arch Lake		Horn Shelter No. 2 (Female)	
	Right	Left	Right	Left
Winging UI1	-	-	-	-
Shoveling UI1	(see depth of shoveling)			
Double Shoveling UI1	-	-	-	-
Interruption grooves UI2	-	-	-	-
Mesial canine ridge UC	-	-	-	-
Premolar odontomes	-	-	-	-
Carabelli cusp UM1	-	-	grade 2	grade 2
Cusp 5 UM1	-	-	-	-
Hypocone UM2	grade 2–3	grade 2–3	grade 2–3	grade 1
Cusp 6 LM1	/	-	/	-
Cusp 7 LM1	-	-	-	-
Distal trigonid crest LM1	/	-	-	-
Protostylid LM1	/	/	-	-
Hypoconulid LM2	/	+	+	+
Y-pattern LM2	/	-	-	+

Key: + = positive (trait is present); - = negative (trait is absent); / = not possible to observe

TABLE 11. DEPTH (MM) OF SHOVELING OF THE UPPER FIRST INCISORS IN TWO PALEOAMERICANS.

Arch Lake			Horn Shelter No. 2 (Female)		
Right	Left	Mean	Right	Left	Mean
0.57	0.83	0.70	0.85	0.79	0.82

leoamericans are given in table 11; and those of comparative recent samples for females and males are presented in tables 12 and 13. Accepting that both Paleoamericans are females, figure 16 shows the mean depth of shoveling between right and left first incisors for female samples. With shoveling depths of 0.70 mm and 0.80 mm respectively, Arch Lake and Horn Shelter No. 2 dentitions have much less shoveling, indicating dissimilarity to the recent Native American samples.

TABLE 12. DEPTH (MM) OF SHOVELING IN UPPER FIRST INCISORS OF FEMALES FROM NORTH AND SOUTH AMERICA.

Sample Name	Right				Left		
	N	Mean	s. d.		N	Mean	s. d.
Alaska Eskimos	12	1.08	0.245		7	0.95	0.318
Aleuts	5	0.76	0.257		5	0.86	0.327
Subarctic	5	0.89	0.378		3	0.79	0.420
California	9	1.24	0.315		9	1.25	0.491
Southwest	16	1.29	0.285		13	1.31	0.328
Plains (North)	15	1.08	0.271		20	1.04	0.251
Plains (South)	18	1.15	0.234		15	1.21	0.336
Northeast (East)	14	1.20	0.323		10	1.29	0.325
Northeast (West)	21	1.09	0.352		19	1.16	0.351
Southeast	37	1.14	0.326		37	1.13	0.387
Peruvians	6	1.19	0.160		4	1.17	0.203

TABLE 13. DEPTH (MM) OF SHOVELING IN UPPER FIRST INCISORS OF MALES FROM NORTH AND SOUTH AMERICA.

Sample Name	Right				Left		
	N	Mean	s. d.		N	Mean	s. d.
Alaska Eskimos	9	1.11	0.267		9	1.2	0.245
Aleuts	13	1.03	0.341		11	0.85	0.213
Subarctic	1	1.01	-		2	1.11	0.460
California	11	1.14	0.251		8	1.15	0.292
Great Basin	4	0.93	0.052		5	0.97	0.280
Southwest	14	1.21	0.384		15	1.16	0.292
Plains (Northern)	18	1.21	0.453		12	1.2	0.380
Plains (Southern)	20	1.13	0.378		19	1.24	0.472
Northeast Woodlands (East)	9	0.95	0.200		7	1.1	0.387
Northeast Woodlands (Midwest)	20	1.11	0.356		17	1.24	0.372
Southeast Woodlands	40	1.12	0.268		32	1.12	0.336
Peruvians	3	1.01	0.185		4	1.24	0.156

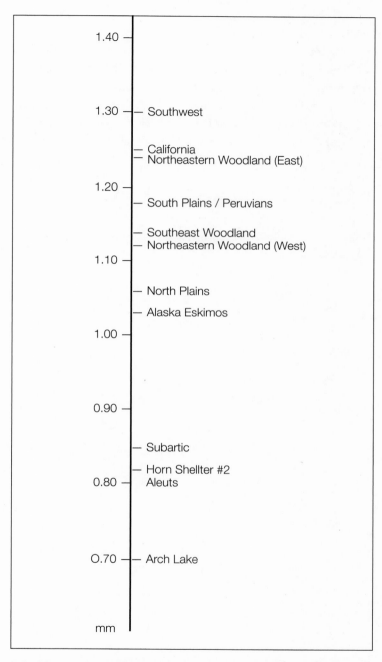

FIGURE 16. Depth of the lingual fossae in upper first incisors of Arch Lake and Horn Shelter No. 2 (female).

Taking into account all of these traits, the two Paleoamericans show several discrete dental traits that are not commensurate with recent Native American groups or Turner's Sinodont group. In other words, they do not show the representative dental traits observed in recent Native Americans.

Mesiodistal and buccolingual crown diameters were also measured for all samples. Measurements for the two Paleoamericans are given in table 14. Measurement means and standard deviations for female and male recent Native

TABLE 14. MESIODISTAL (MD) AND BUCCOLINGUAL (BL) TOOTH CROWN DIAMETERS IN TWO PALEOAMERICAN DENTITIONS.

	Arch Lake			Horn Shelter No. 2 (Female)		
	Right	Left	Mean	Right	Left	Mean
UI1-MD	8.4	8.6	8.5	9.5	9.4	9.5
UI2-MD	7.8	7.7	7.8	8.1	8.1	8.1
UC-MD	8.2	8.3	8.3	8.9	9.1	9.0
UP1-MD	7.5	7.5	7.5	7.7	7.6	7.7
UP2-MD	6.8	6.8	6.8	7.4	7.2	7.3
UM1-MD	11.0	11.0	11.0	12.2	12.4	12.3
UM2-MD	10.7	10.8	10.7	11.3	11.9	11.6
UM3-MD	9.4	8.4	8.9	-	-	-
LI1-MD	-	-	-	-	6.4	6.4
LI2-MD	-	-	-	6.8	7.0	6.9
LC-MD	-	-	-	7.8	7.8	7.8
LP1-MD	6.6	6.8	6.7	7.4	7.3	7.4
LP2-MD	-	-	-	7.6	7.7	7.7
LM1-MD	11.6	11.1	11.4	13.1	13.1	13.1
LM2-MD	-	11.5	11.5	12.9	12.7	12.8
LM3-MD	-	11.8	11.8	-	-	-
UI1-BL	7.3	6.9	7.1	8.0	8.1	8.1
UI2-BL	6.3	6.2	6.3	7.5	7.1	7.3
UC-BL	8.3	8.3	8.3	8.2	8.1	8.2
UP1-BL	9.4	9.8	9.6	10.5	10.6	10.6
UP2-BL	9.5	9.5	9.5	10.0	-	10.0
UM1-BL	12.1	11.8	11.9	12.8	12.9	12.9
UM2-BL	12.0	12.2	12.1	12.8	12.9	12.8
UM3-BL	10.7	10.8	10.8	-	-	-
LI1-BL	-	-	-	-	6.2	6.2
LI2-BL	-	-	-	6.8	6.8	6.8
LC-BL	-	-	-	7.4	7.4	7.4
LP1-BL	7.9	8.2	8.1	7.5	7.8	7.7
LP2-BL	-	-	-	8.7	8.9	8.8
LM1-BL	10.8	10.8	10.8	11.8	11.8	11.8
LM2-BL	-	11.0	11.0	11.3	11.2	11.2
LM3-BL	-	10.3	10.3	-	-	-

American samples are given in tables 15 and 16. Although the juvenile from Horn Shelter No. 2 is probably a female, this individual has large teeth, even in comparison with the recent male samples. As a preliminary analysis, Mahalanobis generalized distances were calculated to estimate the relationship between the two Paleoamericans and recent Native American samples. Group average clustering technique was applied to Mahalanobis generalized distances based on the variance/covariance matrix calculated from the recent female samples pooled. The intergroup similarity/dissimilarity was visualized by applying unweighted pair-group cluster analysis (group average method; fig. 17). The initial split, indicating the greatest dissimilarity, is between a branch of Horn Shelter No. 2 and all of the remaining series. The Arch Lake skull is the next to separate from the remaining recent Native American samples. The reliability of the clustering was tested using a bootstrap method (with 5,000 repeats). The bootstrap probabilities indicate that the two Paleoamerican samples show distinctive dental metrics relative to the recent samples.

The results obtained in the metric and nonmetric dental features suggest that the two Paleoamerican samples do not show close affinities to any recent Native Americans.

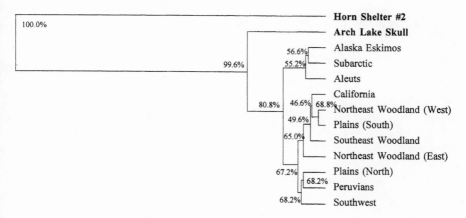

FIGURE 17. Diagram of relationships based on a group average cluster analysis of Mahalanobis generalized distances.

TABLE 15. MESIODISTAL AND BUCCOLINGUAL CROWN DIAMETERS (MM) OF RECENT REGIONAL SAMPLES FROM NORTH AND SOUTH AMERICA (FEMALES).

	Alaska Eskimos			Aleuts			Subarctic			California		
	N	Mean	s.d.	N	Mean	s.d.	N	Mean	s.d.	N	Mean	s.d.
UI1-MD	8	8.76	0.363	6	8.76	0.810	4	8.39	0.240	8	8.96	0.491
UI2-MD	12	7.52	0.409	9	7.67	0.870	4	7.45	0.341	9	7.34	0.749
UC-MD	50	7.96	0.357	14	8.10	0.490	12	8.07	0.313	13	8.08	0.356
UP1-MD	54	7.20	0.390	14	7.30	0.544	14	7.19	0.411	17	7.42	0.379
UP2-MD	53	6.86	0.424	13	6.97	0.572	13	6.78	0.427	17	6.93	0.345
UM1-MD	72	10.79	0.484	18	10.70	0.463	18	10.85	0.507	29	10.88	0.387
UM2-MD	72	10.22	0.521	18	10.01	0.566	18	10.17	0.639	30	10.35	0.465
UM3-MD	58	9.27	0.778	14	8.94	0.671	14	9.37	0.665	18	9.29	0.832
LI1-MD	9	5.40	0.235	6	5.40	0.348	5	5.61	0.115	4	5.77	0.214
LI2-MD	20	6.12	0.292	10	6.22	0.441	7	6.33	0.407	4	6.75	0.315
LC-MD	37	7.01	0.433	11	7.11	0.374	12	7.22	0.274	5	6.98	0.635
LP1-MD	45	6.97	0.318	12	7.01	0.439	9	7.15	0.348	12	7.13	0.433
LP2-MD	48	6.89	0.401	13	7.05	0.594	11	7.04	0.384	11	7.22	0.571
LM1-MD	57	11.64	0.594	20	11.92	0.515	15	11.89	0.663	16	11.90	0.635
LM2-MD	57	11.21	0.686	20	11.12	0.604	16	11.23	0.543	16	11.29	0.689
LM3-MD	43	11.10	1.020	18	11.26	0.792	8	10.98	0.524	10	10.68	0.927
UI1-BL	9	7.38	0.382	6	7.06	0.323	5	7.06	0.200	8	7.34	0.378
UI2-BL	12	6.71	0.467	9	6.69	0.381	5	6.61	0.333	9	6.52	0.514
UC-BL	50	7.90	0.438	14	8.10	0.524	12	8.04	0.488	12	8.20	0.301
UP1-BL	53	9.21	0.541	13	9.04	0.578	14	9.09	0.647	17	9.63	0.319
UP2-BL	54	8.98	0.605	13	9.05	0.651	13	8.89	0.711	17	9.35	0.535
UM1-BL	71	11.74	0.459	17	11.58	0.571	18	11.61	0.514	29	11.93	0.574
UM2-BL	70	11.51	0.600	18	11.31	0.532	18	11.25	0.599	30	11.56	0.521
UM3-BL	58	10.64	0.731	14	10.29	0.696	14	10.39	0.649	18	10.63	0.621
LI1-BL	9	5.66	0.141	5	5.56	0.312	5	5.38	0.255	4	5.88	0.451
LI2-BL	20	6.16	0.248	12	6.11	0.263	6	6.01	0.125	5	6.38	0.287
LC-BL	37	7.48	0.504	12	7.27	0.569	12	7.53	0.328	12	7.40	0.333
LP1-BL	44	7.65	0.413	12	7.44	0.441	9	7.72	0.576	11	7.81	0.366
LP2-BL	48	8.07	0.470	14	8.34	0.859	11	7.92	0.474	11	7.97	0.594
LM1-BL	57	11.16	0.553	20	10.84	0.464	16	10.92	0.340	16	11.02	0.520
LM2-BL	57	10.63	0.528	20	10.36	0.523	16	10.50	0.409	16	10.43	0.494
LM3-BL	43	10.14	0.761	17	10.18	0.497	8	10.00	0.377	10	9.95	0.848

TABLE 15 (cont.).

	Southwest			Plains (Northern)			Plains (Southern)			Northeast (East)		
	N	Mean	s. d.	N	Mean	s. d.	N	Mean	s. d.	N	Mean	s. d.
UI1-MD	12	8.84	0.264	20	8.65	0.400	15	8.92	0.348	13	8.98	0.579
UI2-MD	14	7.70	0.437	21	7.30	0.446	21	7.46	0.443	13	7.67	0.681
UC-MD	17	8.07	0.352	30	8.10	0.341	23	8.14	0.429	19	8.01	0.292
UP1-MD	19	7.47	0.443	36	7.20	0.351	26	7.39	0.370	18	7.31	0.335
UP2-MD	18	7.23	0.430	36	6.99	0.393	26	7.21	0.465	17	7.18	0.391
UM1-MD	22	10.94	0.670	40	10.66	0.426	31	11.01	0.447	21	10.93	0.523
UM2-MD	26	10.19	0.558	39	10.20	0.541	30	10.28	0.549	20	10.37	0.516
UM3-MD	17	9.19	0.821	27	8.89	0.575	17	9.33	0.636	15	9.00	0.711
LI1-MD	7	5.49	0.221	22	5.55	0.346	20	5.55	0.250	12	5.64	0.290
LI2-MD	12	6.20	0.380	27	6.28	0.321	24	6.39	0.276	14	6.39	0.347
LC-MD	16	7.05	0.326	30	7.11	0.302	25	7.24	0.349	18	7.12	0.391
LP1-MD	17	7.06	0.342	33	6.97	0.299	26	7.17	0.348	18	7.15	0.405
LP2-MD	17	7.28	0.398	29	7.19	0.400	26	7.47	0.435	19	7.43	0.442
LM1-MD	16	11.57	0.519	34	11.48	0.446	29	11.70	0.554	12	11.64	0.486
LM2-MD	17	11.20	0.795	33	11.01	0.480	27	11.09	0.645	20	11.44	0.691
LM3-MD	10	11.05	0.904	26	10.40	0.665	27	11.07	0.632	19	11.25	0.911
UI1-BL	12	7.52	0.326	20	7.48	0.441	15	7.38	0.461	13	7.45	0.363
UI2-BL	14	6.85	0.341	21	6.74	0.377	21	6.74	0.536	13	6.71	0.583
UC-BL	15	8.40	0.486	30	8.39	0.430	23	8.08	0.521	19	8.31	0.443
UP1-BL	19	9.40	0.489	35	9.40	0.498	26	9.50	0.419	18	9.71	0.484
UP2-BL	18	9.37	0.409	35	9.27	0.509	25	9.38	0.502	17	9.63	0.659
UM1-BL	23	11.50	0.462	40	11.78	0.459	31	11.63	0.383	21	11.84	0.577
UM2-BL	25	11.22	0.565	39	11.65	0.538	30	11.41	0.530	20	11.67	0.594
UM3-BL	17	10.76	0.689	27	10.60	0.705	17	10.89	0.523	15	10.92	0.740
LI1-BL	6	5.66	0.356	23	5.76	0.236	20	5.77	0.256	12	5.73	0.512
LI2-BL	12	6.12	0.407	28	6.23	0.302	24	6.23	0.317	14	6.20	0.476
LC-BL	16	7.60	0.358	30	7.53	0.487	25	7.60	0.435	18	7.75	0.485
LP1-BL	17	7.80	0.415	33	7.85	0.482	26	7.91	0.516	18	8.09	0.542
LP2-BL	17	8.33	0.452	29	8.14	0.510	26	8.43	0.428	19	8.55	0.521
LM1-BL	15	10.70	0.402	34	10.89	0.426	29	10.97	0.488	13	10.90	0.476
LM2-BL	17	10.41	0.541	33	10.34	0.401	27	10.47	0.555	20	10.59	0.601
LM3-BL	10	10.28	0.809	26	9.81	0.800	27	10.43	0.620	19	10.45	0.748

TABLE 15 (cont.).

	Northeast (Midwest)			Southeast			Peruvians		
	N	Mean	s. d.	N	Mean	s. d.	N	Mean	s. d.
UI1-MD	17	8.82	0.407	32	8.79	0.609	6	8.97	0.588
UI2-MD	19	7.62	0.459	32	7.29	0.702	8	7.07	0.260
UC-MD	22	8.15	0.323	40	8.09	0.443	19	8.27	0.385
UP1-MD	25	7.43	0.312	40	7.26	0.505	38	7.40	0.366
UP2-MD	24	7.10	0.403	41	7.11	0.529	35	7.23	0.442
UM1-MD	28	10.81	0.425	44	11.03	0.624	58	11.00	0.555
UM2-MD	28	10.19	0.552	44	10.18	0.635	54	10.24	0.601
UM3-MD	18	9.53	0.539	35	9.39	0.748	23	9.00	0.773
LI1-MD	18	5.52	0.237	32	5.52	0.463	5	5.67	0.428
LI2-MD	21	6.47	0.405	36	6.27	0.482	11	6.26	0.378
LC-MD	23	6.97	0.286	37	6.97	0.351	13	7.17	0.346
LP1-MD	26	7.17	0.363	43	7.07	0.521	19	6.97	0.418
LP2-MD	26	7.48	0.484	41	7.35	0.540	15	7.45	0.500
LM1-MD	27	11.63	0.510	33	11.61	0.578	18	11.72	0.578
LM2-MD	26	11.22	0.738	37	11.14	0.762	16	10.88	0.739
LM3-MD	18	11.25	0.813	35	11.12	0.813	8	10.32	0.460
UI1-BL	17	6.99	0.392	32	7.08	0.427	5	7.76	0.510
UI2-BL	19	6.70	0.346	32	6.45	0.546	8	6.77	0.706
UC-BL	22	8.15	0.464	40	8.11	0.535	19	8.23	0.596
UP1-BL	25	9.63	0.544	39	9.42	0.624	38	9.49	0.523
UP2-BL	24	9.43	0.566	40	9.25	0.556	34	9.23	0.507
UM1-BL	28	11.65	0.495	44	11.64	0.531	58	11.66	0.544
UM2-BL	28	11.46	0.643	44	11.37	0.628	54	11.34	0.555
UM3-BL	18	10.92	0.614	35	10.75	0.618	23	10.61	0.626
LI1-BL	18	5.47	0.337	32	5.46	0.409	5	5.79	0.316
LI2-BL	21	5.94	0.350	36	6.00	0.431	11	6.13	0.275
LC-BL	23	7.26	0.354	37	7.28	0.457	13	7.63	0.379
LP1-BL	26	7.99	0.497	43	7.74	0.470	19	7.81	0.521
LP2-BL	26	8.50	0.506	41	8.28	0.553	15	8.39	0.564
LM1-BL	27	10.95	0.416	33	10.88	0.537	18	10.87	0.591
LM2-BL	26	10.45	0.547	37	10.33	0.599	16	10.16	0.598
LM3-BL	18	10.42	0.578	35	9.99	0.605	8	9.70	0.579

TABLE 16. MESIODISTAL AND BUCCOLINGUAL CROWN DIAMETERS (MM) OF RECENT REGIONAL SAMPLES FROM NORTH AND SOUTH AMERICA (MALES).

	Alaska Eskimos			Aleuts			Subarctic			California		
	N	Mean	s.d.	N	Mean	s.d.	N	Mean	s.d.	N	Mean	s.d.
UI1-MD	13	9.20	0.675	18	8.72	0.491	3	8.96	0.685	15	9.22	0.395
UI2-MD	18	7.70	0.567	19	7.56	0.395	6	7.63	0.298	13	7.63	0.343
UC-MD	45	8.20	0.350	27	8.46	0.421	12	8.26	0.370	27	8.55	0.326
UP1-MD	44	7.49	0.421	31	7.63	0.441	14	7.25	0.289	28	7.72	0.316
UP2-MD	39	6.93	0.386	28	7.15	0.301	16	7.19	0.428	26	7.44	0.450
UM1-MD	61	11.16	0.514	32	11.06	0.578	20	11.19	0.484	32	11.47	0.620
UM2-MD	61	10.64	0.586	33	10.34	0.551	24	10.56	0.588	33	11.07	0.580
UM3-MD	40	9.57	0.807	19	9.79	1.066	18	9.57	0.909	24	9.78	0.878
LI1-MD	16	5.58	0.362	15	5.61	0.281	4	5.42	0.082	9	5.87	0.331
LI2-MD	21	6.44	0.318	18	6.43	0.398	7	6.35	0.264	10	6.84	0.410
LC-MD	28	7.49	0.461	26	7.66	0.427	16	7.34	0.329	13	7.45	0.438
LP1-MD	36	7.20	0.340	27	7.21	0.372	15	7.00	0.283	19	7.28	0.391
LP2-MD	35	7.06	0.267	30	7.39	0.448	16	7.05	0.376	19	7.41	0.334
LM1-MD	45	12.08	0.571	36	12.26	0.585	18	12.14	0.524	24	12.48	0.473
LM2-MD	46	11.75	0.647	33	11.56	0.699	21	11.57	0.563	25	12.00	0.484
LM3-MD	37	11.57	0.959	26	11.74	0.852	19	11.20	0.659	22	11.67	0.892
UI1-BL	15	7.56	0.354	18	7.51	0.440	3	7.57	0.201	15	7.49	0.441
UI2-BL	18	7.04	0.627	20	7.07	0.470	7	7.05	0.480	13	6.74	0.359
UC-BL	45	8.55	0.480	26	8.47	0.477	12	8.23	0.464	27	8.68	0.456
UP1-BL	44	9.53	0.675	31	9.44	0.533	15	9.01	0.386	27	9.86	0.470
UP2-BL	38	9.31	0.539	28	9.36	0.455	14	9.08	0.409	26	9.71	0.549
UM1-BL	60	12.20	0.499	32	12.02	0.465	19	11.93	0.486	32	12.38	0.516
UM2-BL	61	12.12	0.708	32	11.73	0.598	24	11.81	0.495	33	12.25	0.586
UM3-BL	40	11.05	0.775	19	11.26	0.903	18	10.92	0.849	24	11.07	0.640
LI1-BL	16	6.16	0.443	15	5.70	0.388	5	5.68	0.264	9	5.83	0.280
LI2-BL	21	6.51	0.425	18	6.28	0.481	7	6.17	0.295	10	6.28	0.462
LC-BL	29	8.17	0.475	26	8.01	0.467	15	7.80	0.446	13	7.80	0.449
LP1-BL	36	8.10	0.502	27	7.93	0.417	14	7.80	0.378	19	8.29	0.376
LP2-BL	34	8.40	0.536	30	8.40	0.457	16	7.91	0.352	18	8.44	0.357
LM1-BL	45	11.54	0.436	35	11.21	0.432	20	11.13	0.588	23	11.37	0.505
LM2-BL	46	11.11	0.618	32	10.87	0.480	21	10.76	0.513	25	11.06	0.492
LM3-BL	37	10.61	0.647	25	10.63	0.635	20	10.29	0.681	22	10.58	0.598

46

TABLE 16 (cont.).

	Great Basin			Southwest			Plains (Northern)			Plains (Southern)		
	N	Mean	s.d.	N	Mean	s.d.	N	Mean	s.d.	N	Mean	s.d.
UI1-MD	5	8.82	0.378	14	8.93	0.464	25	8.92	0.495	24	9.13	0.434
UI2-MD	6	7.68	0.446	17	7.39	0.400	29	7.70	0.512	27	7.82	0.523
UC-MD	8	8.32	0.433	19	8.32	0.460	32	8.34	0.445	33	8.40	0.485
UP1-MD	9	7.59	0.362	22	7.40	0.365	36	7.34	0.461	38	7.56	0.449
UP2-MD	8	7.34	0.278	24	7.13	0.377	34	7.10	0.418	38	7.27	0.487
UM1-MD	12	11.26	0.488	25	11.10	0.479	38	11.03	0.599	45	11.33	0.574
UM2-MD	12	10.68	0.871	24	10.50	0.499	38	10.55	0.530	43	10.51	0.645
UM3-MD	8	9.39	0.882	17	9.21	0.728	31	9.51	0.750	38	9.50	0.761
LI1-MD	3	5.66	0.074	8	5.44	0.235	21	5.55	0.309	17	5.60	0.317
LI2-MD	4	6.51	0.281	9	6.10	0.232	27	6.35	0.418	24	6.46	0.457
LC-MD	6	7.45	0.508	11	7.41	0.331	31	7.53	0.370	30	7.40	0.496
LP1-MD	8	7.37	0.598	17	7.02	0.356	31	7.17	0.442	37	7.23	0.506
LP2-MD	9	7.71	0.530	15	7.13	0.367	33	7.38	0.534	37	7.49	0.567
LM1-MD	12	12.26	0.538	20	11.82	0.422	36	11.95	0.634	39	12.07	0.648
LM2-MD	12	11.71	0.786	20	11.16	0.550	35	11.56	0.657	40	11.59	0.797
LM3-MD	8	11.17	0.848	20	10.87	0.807	28	11.32	0.841	35	11.34	0.891
UI1-BL	5	7.52	0.335	14	7.63	0.297	27	7.65	0.535	24	7.81	0.455
UI2-BL	6	7.12	0.646	17	6.65	0.342	30	6.88	0.401	27	6.97	0.448
UC-BL	8	8.74	0.602	19	8.55	0.577	33	8.85	0.519	33	8.74	0.563
UP1-BL	9	9.83	0.484	21	9.42	0.453	36	9.69	0.557	38	9.76	0.578
UP2-BL	8	9.55	0.679	23	9.29	0.517	32	9.56	0.572	37	9.54	0.688
UM1-BL	12	12.13	0.464	24	11.85	0.494	38	12.06	0.470	46	12.06	0.634
UM2-BL	12	12.06	0.840	24	11.73	0.585	38	12.07	0.594	44	11.95	0.790
UM3-BL	8	11.26	0.691	17	11.04	0.805	31	11.24	0.618	39	11.21	0.833
LI1-BL	3	5.90	0.255	8	5.91	0.291	24	5.94	0.411	18	6.04	0.440
LI2-BL	4	6.23	0.204	10	6.19	0.416	28	6.40	0.459	23	6.39	0.460
LC-BL	6	8.13	0.603	11	8.01	0.415	31	8.20	0.471	30	8.17	0.558
LP1-BL	8	8.05	0.359	17	7.87	0.579	31	8.14	0.471	36	8.15	0.574
LP2-BL	9	8.66	0.439	15	8.30	0.459	33	8.54	0.478	37	8.47	0.546
LM1-BL	10	11.29	0.344	20	11.04	0.575	36	11.22	0.474	36	11.27	0.642
LM2-BL	12	10.95	0.562	18	10.59	0.458	34	10.83	0.463	40	10.74	0.594
LM3-BL	8	10.61	0.543	19	10.28	0.488	28	10.58	0.529	35	10.53	0.699

TABLE 16 (cont.).

	Northeast (East)			Northeast (West)			Southeast			Peruvians		
	N	Mean	s.d.	N	Mean	s.d.	N	Mean	s.d.	N	Mean	s.d.
UI1-MD	9	9.19	0.277	30	9.11	0.404	36	8.98	0.421	5	9.28	0.393
UI2-MD	11	7.71	0.513	33	7.76	0.529	39	7.69	0.651	8	7.9	0.621
UC-MD	16	8.57	0.444	38	8.49	0.399	45	8.40	0.414	19	8.45	0.416
UP1-MD	17	7.58	0.445	43	7.58	0.475	45	7.50	0.457	40	7.72	0.356
UP2-MD	17	7.38	0.511	43	7.23	0.431	45	7.32	0.511	33	7.43	0.394
UM1-MD	18	11.58	0.475	46	11.25	0.469	47	11.27	0.573	43	11.31	0.564
UM2-MD	18	10.74	0.501	45	10.75	0.549	48	10.66	0.752	44	10.77	0.734
UM3-MD	16	9.57	0.992	39	9.54	0.806	35	9.40	0.930	30	9.40	0.894
LI1-MD	9	5.55	0.163	32	5.70	0.307	24	5.61	0.323	5	5.90	0.252
LI2-MD	11	6.34	0.276	36	6.52	0.345	32	6.45	0.456	12	6.37	0.226
LC-MD	11	7.49	0.354	41	7.51	0.355	34	7.49	0.489	14	7.48	0.314
LP1-MD	16	7.54	0.452	42	7.39	0.386	39	7.26	0.399	19	7.42	0.377
LP2-MD	16	7.65	0.584	42	7.59	0.512	41	7.52	0.489	22	7.73	0.531
LM1-MD	19	12.24	0.374	43	11.97	0.432	37	12.02	0.475	29	12.22	0.472
LM2-MD	14	11.92	0.664	43	11.81	0.723	34	11.66	0.581	28	11.98	0.872
LM3-MD	16	11.95	0.820	41	11.67	0.860	36	11.21	0.961	18	11.23	0.798
UI1-BL	9	7.47	0.403	30	7.57	0.449	36	7.49	0.442	5	7.97	0.469
UI2-BL	11	6.83	0.556	33	6.91	0.517	39	6.70	0.515	8	7.20	0.285
UC-BL	16	8.74	0.443	38	8.78	0.541	45	8.58	0.528	19	8.77	0.506
UP1-BL	17	10.06	0.576	42	9.89	0.604	45	9.74	0.582	40	9.86	0.436
UP2-BL	17	9.78	0.440	43	9.68	0.660	45	9.74	0.570	33	9.73	0.516
UM1-BL	18	12.40	0.433	46	12.11	0.560	47	12.14	0.485	43	12.11	0.525
UM2-BL	18	12.21	0.426	45	12.04	0.619	48	12.16	0.696	43	12.03	0.691
UM3-BL	16	11.49	0.816	38	11.31	0.798	35	11.12	0.906	30	11.15	1.006
LI1-BL	9	5.99	0.245	35	5.96	0.410	24	5.73	0.340	5	6.16	0.305
LI2-BL	10	6.32	0.332	36	6.33	0.340	32	6.13	0.330	11	6.45	0.385
LC-BL	11	8.02	0.348	41	8.05	0.476	34	7.84	0.463	13	8.15	0.522
LP1-BL	16	8.19	0.513	42	8.32	0.493	39	8.07	0.520	19	8.27	0.510
LP2-BL	16	8.68	0.519	42	8.69	0.566	41	8.44	0.499	21	8.69	0.780
LM1-BL	18	11.44	0.407	43	11.32	0.537	36	11.21	0.468	29	11.37	0.501
LM2-BL	14	11.11	0.380	43	11.01	0.541	34	10.70	0.433	28	10.95	0.617
LM3-BL	16	10.99	0.543	41	10.80	0.633	36	10.37	0.709	18	10.56	0.764

Post-cranial Measurements

ETRIC analysis of post-cranial bones has focused on the femur and tibia, and interpretation emphasizes the biomechanical, behavioral, and adaptive implications of size and shape. Basic post-cranial dimensions could be obtained on the somewhat fragmented remains of the Arch Lake skeleton. Table 17 presents femur and tibia measurements and indices for Arch Lake and three comparative samples: Pecos Pueblo, Coalescent Arikara, and American whites from the Terry anatomical collection. Pecos data were obtained from E. Hooton's data cards (Weisensee 2001), the Coalescent Arikara from Cole (1994), and the Terry data were provided by S. D. Ousley, Department of Anthropology, National Museum of Natural History. Arch Lake variables were tested against the comparative samples using a t-test comparing an individual to a sample, as recommended by Jolicoeur (1999). The platymeric, pilastric, and cnemic indices are computed as the anterior-posterior (a-p) dimension divided by the medial-lateral (m-1), so that a value > 1.0 describes a-p elongation, and values < 1.0 reflect m-1 expansion. Robusticity indices at the level of the femur midshaft, subtrochanter, and nutrient foramen of the tibia are computed as (a-p + m-1)/maximum bone length x 100. The humerus was treated in the manner described by Collier (1989) in order to use his comparative data. Humerus robusticity is computed as circumference/humerus length. Arch Lake's values and data from Collier (1989) are presented in table 18.

The platymeric, pilastric and cnemic indices presented in table 17 describe the shape of the diaphyses at the subtrochanteric and midshaft levels of the femur and at the nutrient foramen of the tibia. The most striking feature of Arch Lake is the absence of platymeria, a feature commonly seen in recent Native Americans (Gill 1995). Arch Lake differs significantly from the two Native American samples in table 17 but is similar to American whites. Arch Lake is also somewhat more robust in the subtrochanteric region than other groups, but not significantly so.

TABLE 17. FEMUR AND TIBIA DIMENSIONS (MM) AND SELECTED INDICES FOR ARCH LAKE COMPARED TO TWO FEMALE NATIVE AMERICAN SAMPLES AND AMERICAN WHITES.

Variable	Arch Lake	Pecos Pueblo			Coalescent Arikara+			American Whites		
		N	Mean	s. d.	N	Mean	s.d.	N	Mean	s. d.
Fem max length	447	109	394.51**	17.49	192	415.20	17.30	56	426.50	22.63
Subtroch a-p	29	109	21.45**	1.54	193	22.40**	1.80	56	26.23	2.03
Subtroch m-l	31	109	29.26	1.91	193	32.10	1.90	56	27.75	2.06
Midshaft a-p	28	109	25.12	2.08	194	25.90	2.10	56	26.45	2.12
Midshaft m-l	26	109	23.11*	1.41	194	24.40	1.50	56	26.07	1.92
Platymeric index	0.94	109	0.74**	0.66	193	0.72**	0.06	56	0.95	0.09
Pilastric index	1.08	109	1.09	0.10	194	1.06	0.09	56	1.02	0.09
Subtroch robusticity	13.42	109	12.87	0.72	192	12.88	0.68	56	12.66	0.78
Midshaft robusticity	12.08	109	12.24	0.71	192	12.11	0.65	56	12.31	0.79
Tibia max length	359	74	323.46*	15.34	192	348.00	16.10	56	343.86	19.42
Nut foramen a-p	32	74	30.17	1.95	194	31.20	2.50	56	29.13	1.92
Nut foramen m-l	21	74	19.50	1.58	194	22.00	2.20	56	23.00	2.05
Cnemic index	1.52	74	1.56	0.15	194	1.43	0.15	56	1.27*	0.11
Nut for robusticity	14.76	74	15.38	0.93	192	15.30	1.03	56	15.16	0.93
Crural Index	80.31	67	82.48	2.21	174	83.88	2.02	54	80.44	2.64

* $P < 0.05$
** $P < 0.01$
+ From Cole 1994, from raw data.

TABLE 18. ARCH LAKE HUMERUS MEASUREMENTS (MM) COMPARED TO
WORLD POPULATION SAMPLES.

Group	Humerus Length		Humerus Circumference		Robusticity	
	Mean	s. d.	Mean	s. d.	Mean	s. d.
Arch Lake	284	-	64	-	22.54	-
Australian	306.4	17.6	52.3	4.8	17.1*	1.6
Whaling Eskimo	284.4	17.4	55.8	4.7	19.6	1.4
Riverine Eskimo	285.7	14.8	53.4*	2.9	18.7*	1.1
American Whites	304.0	14.9	56.2	4.0	18.5*	1.4
Arikara Indians	301.1	13.1	57.1	3.1	19.0*	1.1
Romano-Britons	298.3	13.8	57.1	3.2	19.1*	1.2

* Differs significantly from Arch Lake

Ruff (2000) has argued that femur cross section shape reflects mobility. The argument, briefly, is that mobility imposes anterior-posterior loading on the femur midshaft, which results in remodeling to enable the femur to resist the loading pattern with an a-p elongated femur midshaft. Hence the degree of femur a-p elongation has been taken as an indicator of mobility, even to the point of referring to it as a mobility index (Larsen 1997). Later, Ruff (2000) argued that terrain was the most important variable explaining femur a-p elongation. Wescott (2001) tested and elaborated this hypothesis using North American samples that included a wide range of subsistence types, culture areas, and terrain types. His results provide some support that femur cross sectional morphology reflects activity, as opposed to mobility, and that the effect is greater at the subtrochanteric level than at the midshaft. Shaft robusticity also reflects activity (Collier 1989).

Arch Lake's femur midshaft, expressed by the pilastric index, shows a-p elongation comparable to the two Native American comparative samples and greater than in the nineteenth-century whites. Arch Lake's robusticity exceeds the comparative groups at the subtrochanteric level but is more gracile than the comparative groups at the femur midshaft. The Arch Lake tibia is also gracile and, along with Pecos, has a-p elongation exceeding that of the other groups.

The humerus picture is quite different. Its maximum length is short, below average for all comparative groups, and its circumference is large, above average for all comparative groups. Neither dimension differs significantly from the comparative groups, except for Riverine Eskimos, but length and circumference taken together present a picture of a very robust humerus that differs from nearly all comparative groups. Given her nitrogen stable isotope value indicating a diet dominated by high quality protein, a strenuous recurrent task that might account for this robusticity was working bison hides.

Arch Lake femur and tibia lengths exceed the means of all three comparative groups. In the case of Pecos, the difference is significant. Arch Lake femur and tibia lengths do not differ significantly from Coalescent Arikara or Terry collection whites. The combined length of the femur and tibia also exceeds means for Neanderthals, European Upper Paleolithic, Mesolithic, and recent humans. The crural index (tibia length/femur length X 100) of Arch Lake is lower than seen in Coalescent Arikara or Pecos and roughly equal to the white sample. It is not as low as seen in cold-adapted populations, such as Eskimos, but not as high as seen in the European Upper Paleolithic or Mesolithic (Holliday 1999).

Estimation of stature, while not necessary in terms of understanding bone dimensions, is useful to provide an indication of how early remains relate to modern people using a widely understood measure. It is difficult to know which formula to use, because we do not know the body proportions of the Arch Lake woman. Many modern Native North Americans have relatively long legs (Jantz et al. 2002), and some, such as the Coalescent Arikara, have relatively high crural indices (Hall et al. 2004). However, many Mexican tribes have relatively short legs (Faulhaber 1970). Ruff has argued that stature formulae applied to fossils should be matched as closely as possible to the limb proportions of the fossil. Following that advice would lead to choosing the formulae of whites, the group to which the crural index of Arch Lake is most similar. Classical calibration can also be used, regressing bone length on stature and then solving for stature, which yields unbiased estimates (Konigsberg et al. 1998).

Table 19 presents stature estimations using various approaches. Using whites assumes the Arch Lake woman has relatively short legs in relation to stature, while using blacks assumes she has relatively long legs relative to stature. Arch Lake's crural index suggests her proportions may have been more like those of whites. The femur exhibits lower positive allometry with stature, so it is likely the more reliable. Tibia estimates are lower because of the relatively short tibia of Arch Lake. The humerus yields the lowest estimates, indicating that it is relatively short compared to the legs. The estimates in table 19 indicate that the Arch Lake female could have been quite tall. The largest estimate of 166.5 cm (5 feet 5½ inches) places her approximately equal to modern American whites. If one assumes relatively long legs; however, she would be somewhat shorter.

TABLE 19. ESTIMATION OF STATURE OF THE ARCH LAKE FEMALE USING INVERSE CALIBRATION (TROTTER AND GLESER 1952) AND CLASSICAL CALIBRATION.

Source	Femur	Tibia	Humerus
Trotter and Gleser Whites	164.51 ± 3.72 (64.8")	162.74 ± 3.66 (64.1")	153.36 ± 4.45 (60.4")
Trotter and Gleser Blacks	161.68 ± 3.41 (63.7")	158.16 ± 3.70 (62.3")	152.14 ± 4.25 (59.9")
Classical Calibration Whites	166.50 ± 4.76 (65.6")	163.41 ± 5.33 (64.3")	---
Classical Calibration Blacks	162.16 ± 3.79 (63.8")	156.86 ± 4.53 (61.8")	---

Burial Assemblage

T HE Arch Lake burial included a total of twenty-two artifacts: nineteen drilled beads, a unifacial stone tool, apparently pulverized red ocher, and a probable bone tool. None of these material culture items is culturally diagnostic.

Flake tool

A resharpened flake tool was found near the woman's waist on her left side, associated with a dense concentration of red ocher (see 1 in fig. 4, color plate). Excavation photos show red pigment adhering to the tool upon discovery. Subsequent cleaning removed most of this, but remnants are visible near flake scar terminations.

The uniface is made of a fine-grained, cryptocrystalline chert of variegated hues of light brown and cream. The material is identified as Edwards chert on the basis of visual and microscopic examination and direct comparison with geologic samples using short- and long-wave ultra-violet fluorescence. The Arch Lake tool fluoresces in similar shades of orange-yellow to the comparative Edwards samples.

Maximum dimensions of this uniface are 51.2 mm long by 25.3 mm wide by 4.9 mm thick. Both lateral edges exhibit extensive, continuous, and regular retouch that extends from the tool margins nearly 2 to 4 mm onto the dorsal face of the flake (fig. 18). One retouched edge is slightly convex with relatively steep edge angles (> 45°) and measures nearly 42 mm long; it terminates in a narrow, truncated bit (4.9 mm wide by nearly 1.3 mm thick) formed at the proximal end of the flake. The opposite lateral margin is gently concave and measures nearly 29.5 mm in length (fig. 19, color plate).

The platform and bulb of percussion are missing. The bulb's former position is overprinted by a flake scar directed at an approximate angle of 45° to the long axis of the flake blank. This flake removal eliminated the bulb of percus-

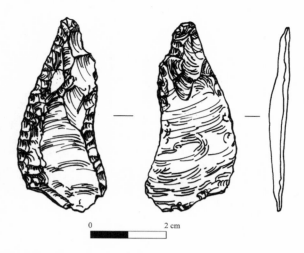

0   2 cm

FIGURE 18. Illustration of the unifacial flake tool associated with the burial.

sion. In turn, invasive tool resharpening on the adjacent dorsal face truncated the proximal end of this bulb-removal flake scar. The distal end of the tool is edge damaged and exhibits snap breaks.

Observations of tool morphology and low power microscopic use-wear analysis suggest that the tool was used for tasks requiring light duty cutting and scraping. The multiple edge configurations suggest a Swiss Army knife type of use. The narrow, chisel-shaped bit (4.9 mm wide) is reminiscent of the "tranchet tip" (6 mm wide) reported on the distal end of the Windust stemmed biface found with the Buhl burial (Green et al. 1998:449).

The Arch Lake uniface is not a culturally diagnostic artifact. Rather, it fits comfortably within more than one Paleoamerican tradition of stone tool making.

Red Pigment

No consolidated pieces of red ocher were recovered. Rather, the pigment appears to have been placed in the grave in an unconsolidated form, perhaps ground or pulverized. The ocher was most concentrated, vivid in color, and thickest in a vertical section in an area measuring approximately 65 mm by 75 mm by 55 mm deep where the stone tool was found. Concentrated pigment extends from here along the lateral side of the left humerus to midshaft. Some

disturbance to these sediments by bioturbation is evident in the original photograph (fig. 4).

A cross-section profile through the matrix block in the abdominal/lumbar
region shows variably saturated red-to-pink hues permeating the white sand
of the burial pit fill. Areas of diffuse, pink-stained sediment were noted lateral
to the right humerus and femur, suggesting that red pigment may have been
lightly sprinkled on the grave and/or body.

Abundant red ocher covered the adult woman and grave goods buried at
Gordon Creek. At Horn Shelter No. 2, a large red ocher nodule was found
beneath the head of the adult male, but additional red pigment was not spread
over either of the two bodies nor any of the burial goods. Red pigment was
similarly absent from Burial 2 at Wilson-Leonard, nor did it appear to accom-
pany the remains at Buhl.

Talc Beads

Nineteen biconically drilled, disc-shaped beads made of opaque to slightly
translucent white indurated talc were recovered at Arch Lake. Fourteen beads
were found in an arc just above the clavicles, suggesting they were strung in a
necklace. Five additional beads were collected while screening burial fill.

Fifteeen beads were available for study at the Blackwater Draw Museum in
February 2000 (fig. 19, color plate). Three others (described later) were on loan
for material analysis, and one was missing from the collection. Maximum diameters (width) range from 5.1 mm to 7.3 mm (table 20). Bead length (cross-
section thickness) ranges from 1.9 mm to 3.5 mm and averages 2.4 mm. The
lengths of thirteen beads range between 1.9 mm and 2.6 mm, with two other
beads measuring 3.1 mm and 3.5 mm.

The holes are biconical with larger diameters at the surface that narrow
toward the center, generally indicating that the hole was partially drilled from
each face. The range for hole diameters is 1.5 mm to 2.8 mm.

Bead Material Identification and Production

Three beads were analyzed to determine the material from which they had
been carved. They differ from each other in terms of color, shape, and size. The
largest bead is 6.5 mm in diameter and 4.0 mm long with a 2.5 mm to 3 mm

TABLE 20. ARCH LAKE BEAD DIMENSIONS AND CHARACTERISTICS.

Bead	Maximum Dimension (width, plan view)	Hole Diameter Face 1	Hole Diameter Face 2	Color	Additional comments
1	6.2	3.2	1.5	White, opaque	Hole placed off center with polish on its perimeter, no linear striations across face of bead
2	5.7	2.4	2.4	Translucent	Hole maintains consistent diameter
3	7.3	2.9	---	White, opaque	Bi-conically drilled, some polish on hole perimeter; linear striations on face of bead
4	5.7	2.8	2.0	White, opaque	Hole placed off center
5	6.0	2.6	2.6	Slightly translucent	Linear striations on bead exterior oriented perpendicular to bead faces
6	5.3	2.1	2.1	Translucent	Hole centered
7	5.1	2.7	2.3	White, opaque	Holes placed slightly off center and taper inward, clearly bi-conically drilled
8	5.6	2.3	2.3	Translucent	Bi-conically drilled, one hole tapers inward
9	5.3 to 5.8	2.6	2.6	White, opaque	Hole placed off center
10	5.4	2.8	---	White, opaque	Hole tapers inward
11	5.5	2.6	---	White, opaque	Bead has polished exterior and concave faces
12	5.9 to 6.3	2.6	---	White, opaque	Hole centered, no linear striations on faces of bead
13	5.4 to 5.5	2.5	2.4	White, opaque	
14	5.1	2.6	2.5	White, opaque	Hole centered
15	5.4	2.5	2.4	White, opaque	Hole placed off center
16	6.5	2.5	3.0	White, opaque	4 mm long
17	6.0	2.0	3.0	White, opaque	2.5 mm long
18	5.5	2.0	---	Yellowish, slightly translucent	1.5 mm to 2.5 mm long

Note: Measurements in mm. Beads 1–15 were examined in Portales, February 2000; their lengths range from 1.9 mm to 3.5 mm. Beads 16–18 were examined in conjunction with the material analysis.

diameter hole. It is opaque, white in color, and differs from the other eighteen beads in that it is slightly rounded in shape and is faceted across its length. The medium bead is 6.0 mm in diameter, 2.5 mm thick, and has a 2 mm to 3 mm diameter perforation. It is opaque, white in color, and has two flat ends. The smallest bead is 5.5 mm in diameter and 1.5 mm to 2.5 mm thick (slightly irregular), with a 2 mm diameter perforation. It is yellowish in color and slightly translucent. It is also disk shaped with flat ends, although they are not as flat as on the medium-sized bead.

Circular striated grooves are present on the inside surface of the perforations in each of the beads. These grooves occur on all three of the beads analyzed for material identification but are most evident on the large one. The holes tend to be widest at the surface of the beads and narrowest at their center; this feature is most evident in the largest bead. These characteristics indicate that a slender drill was twisted in a circular motion from the two ends of the bead, creating pits that met in the center. Each of the perforations is somewhat asymmetrical, perhaps reflecting wear that resulted from stringing and wearing the beads.

Overall, these three beads are representative of all beads recovered with the burial. Three of the beads are generally similar to the small, translucent one. Twelve are similar to the medium-sized bead. The large bead is unique, as it is the only one that is faceted.

Non-destructive testing was a prerequisite of the elemental analysis. The two methods employed were electron microprobe analysis and X-ray diffraction. Using the microprobe, two areas on each of the three test beads were sampled to determine chemical composition. As carbon is a contaminant that is likely present on each sample, results were generated twice during each analysis; once with carbon included in the result and once with it removed. When carbon was included as a component, it was the most prevalent element by atom percentages in the large and medium beads, and second only to oxygen in the small bead (table 21). Without carbon in the results, oxygen, silicon, and magnesium were the most common elements (table 22). From this initial evaluation, it was evident that the beads were made from the same material: either a magnesium-rich carbonate mineral or a silicate rock with carbon present as a surface contaminant.

When the beads were photographed, a bright green, crystalline substance

TABLE 21. ELECTRON MICROPROBE ANALYSIS OF THREE BEADS WITH CARBON INCLUDED IN THE ANALYSIS.

	Large		Medium		Small	
Element	Sample 1	Sample 2	Sample 1	Sample 2	Sample 1	Sample 2
C	39.37	41.95	39.32	39.04	51.48	46.43
O	42.76	42.50	44.17	44.32	36.93	40.16
Si	9.68	8.41	8.41	8.45	5.92	6.64
Mg	6.69	6.15	6.08	6.10	3.70	4.61
Ca	0.30	0.41	1.49	1.60	1.25	1.50
Fe	0.07	0.03	0.08	0.05	0.13	0.08
Al	1.04	0.49	0.40	0.38	0.49	0.48
K	0.05	0.05	0.04	0.04	0.09	0.06
Cl	0.03	0.01	0.01	0.02	0.01	0.04

Note: Atom percentages are displayed for two repeat tests. Each bead was sampled twice.

TABLE 22. ELECTRON MICROPROBE ANALYSIS OF THREE BEADS WITH CARBON EXCLUDED FROM THE ANALYSIS.

	Large		Medium		Small	
Element	Sample 1	Sample 2	Sample 1	Sample 2	Sample 1	Sample 2
O	60.33	62.44	64.31	64.23	64.67	64.87
Si	22.71	21.50	19.02	18.93	18.90	18.19
Mg	13.52	13.55	11.26	12.32	10.45	11.26
Ca ·	0.67	1.01	3.46	3.43	3.74	3.87
Fe	0.15	0.08	0.31	0.10	0.36	0.20
Al	2.41	1.26	1.42	0.87	1.57	1.32
K	0.12	0.12	0.20	0.08	0.27	0.17
Cl	0.09	0.04	0.03	0.04	0.03	0.12

Note: Atom percentages are displayed for two repeat tests. Each bead was sampled twice.

was noted on the interior walls of the bead holes. Electron microprobe analysis of this substance provided clues to its composition, but definitive identification of its nature and origin was not made. The primary compositional difference between the green areas and the calcite surface areas was the presence of either titanium or a combination of sulfur and barium. When titanium levels were high, silicon levels also increased. Despite the presence of these other elements, calcium is still well represented, as are silicon and oxygen. Magnesium, carbon, aluminum, potassium, manganese, and iron round out the chemical composition of the green, crystalline substance.

Due to the indeterminate results of the electron microprobe analysis, X-ray diffraction was undertaken. X-ray diffraction reveals the structure of an unknown crystal and thus identifies it. The technique is possible because of the regular structure of crystals and the manner in which X-rays interact with electrons (Klein and Hurlbut 1999:2, 32). The diffraction pattern produced is unique to a particular substance (mineral species) because it reflects the fundamental physical properties of its crystal lattice structure (Klug and Alexander 1974:121).

Although an X-ray powder diffractometer was used for this analysis, the usual procedure, which reduces the sample to fine particles, was modified in order to avoid destroying the bead. Instead, the medium-sized bead was set into a hole in an aluminum base so that one of the bead's flat ends was flush with the top of the base and level with the surface on which the powder is usually placed. The analysis then continued in the standard manner.

The X-ray diffraction analysis identified the material from which the beads had been manufactured as talc, $Mg_3Si_4O_{10}(OH)_2$. This result yielded a similar chemical composition to that obtained through the microprobe analysis. As illustrated by table 22, when carbon was assumed to be present as surface contamination, and not a component of the material from which the beads were carved, the primary elements detected matched those that form talc: oxygen, silicon, and magnesium. With evidence from one analytical technique that characterized the chemical composition of the beads and another that identified the crystalline structure of the material from which they were carved, the identification of talc is strongly supported.

The determination of talc was somewhat unexpected because talc is generally a very soft mineral; it is the standard for 1 on the Mohs scale of hardness. The material from which the beads had been carved is harder than this. Under some conditions, however, hardness in talc can vary. Compact, massive, cryptocrystalline talc, sometimes known as steatite, can range in hardness from 1 to 2.5 on the Mohs scale. According to Spence (1940:9) this variant can form either when areas within an ore body do not undergo complete recrystallization during metamorphism or during the hydrothermal alteration of low-iron magnesium silicate minerals. Unfortunately, the term steatite has been used inconsistently in the past, making references to this type of talc somewhat unclear in the literature (Greene 1995:9–10). Indurated talc, which is slaty and less pure than ordinary talc, is also a harder form of this mineral (Dana 1932:678). Steatite is known to be suitable for carving (Spence 1940:9), and in prehistoric North America talc-bearing rocks were carved into dishes, tools, and ceremonial objects (Greene 1995:8).

While talc is found in many areas of the United States, talc-bearing deposits in the western part of the country are dispersed over a wide area with individual sources fairly localized in extent (Greene 1995; Chidester and Worthington 1962). Based solely on proximity to the Arch Lake burial site, the most obvious possible sources of talc are the Hembrillo Canyon deposit in New Mexico (320 km distant), the Alamore district near Van Horn, Texas (400 km distant), and the Llano uplift in Central Texas (560 km distant; Chidester and Worthington 1962).

A determination of the specific source of the talc from which the beads were carved has not been made. The distribution of currently known talc deposits suggests that these artifacts, in finished or unfinished form, possibly traveled a considerable distance from their source to the place of interment with the Arch Lake burial. Alternatively, sources for talc may yet be identified closer to Arch Lake (J. Warnica, pers. comm. 2000, 2006).

Probable Bone Tool

A nonhuman bone found in the burial matrix lying over the woman's ribs, visible in photographs from the original excavation, appears rounded at one end and is thought to have been a tool. It is missing from the collection and was unavailable for study. Discussion here draws on excavation photographs, small fragments of this bone collected from its pedestal in 2000, and observations made by those who examined the object at the Blackwater Draw Museum in previous years.

This large, cylindrical object was lying diagonally across the chest of the Arch Lake woman, above ribs 8, 9, and 10 (see 2 in fig. 4). Later field photographs confirm that the bone was collected before the burial was removed, but its pedestal remained intact when the burial block was cast for transport to Eastern New Mexico University. J. Dickenson verified that this bone had been replaced on its pedestal while the burial block was on exhibit at the Blackwater Draw Museum. She observed that the bone was modified or rounded at one end and appeared to be a tool. The opposite end was obscured by heavy carbonate encrustation.

In 2000 we collected small fragments of this bone still adhering to its pedestal. This sample (AL-18) consists of one large (length nearly 40 mm) and four small pieces, all of which may originally have been part of the same fragment. Zooarchaeologist Lynn Snyder (written communication 2005) identified the largest specimen as a split long bone cortical fragment; the other four pieces are consistent in size and may well refit. All surfaces are heavily weathered/abraded and chalky in appearance, and none of the original cortical surface is preserved. The fragments are unburned, and no traces of deliberate modification remain. These fragments are consistent in size and cortical thickness with large mammals, minimally in the size range of antelope or deer. Excavation photographs show that the tool had a larger diameter than the midshaft of the woman's humerus visible nearby. This contrast suggests the likelihood that the tool was made of bone from a large animal, such as bison. The implement appears quite robust. Further thoughts on its function await its rediscovery and analysis.

Archaeological Comparison of Arch Lake
with Other Early Burials, 9500–10,020 RC yr.

✦

TABLES 23, 24, and 25 provide information on age and sex, possible cultural affiliation, aspects of burial practice, and the nature of grave goods for five primary internments with AMS dates and/or chronostratigraphic ages between 9500 and 10,020 RC yr. These ancient burials are Arch Lake, Horn Shelter No. 2, Wilson-Leonard burial 2, Gordon Creek, and Buhl. This burial sample spans nearly 500 RC yr. and includes six individuals: single interments of four adult females and a double interment of an adolescent female and an adult male (table 23). The geographic distribution of these archaeological sites is shown in figure 20.

General comparisons are also heuristically drawn between this burial assemblage and those accompanying a larger, though younger (6990–8120 RC yr.), cemetery series from the Windover site in Florida. Eighty-two of the 169 individuals recovered at Windover had burial artifacts, 63 did not, and 24 were indeterminate for artifact association due to commingled deposits. Of the 82, 23 were female, 29 were male, and 30 were of indeterminate sex (Hamlin 2001).

Burial Position

The Arch Lake woman is unique within the older series with regard to positioning: on her back with legs extended. The other interments are flexed burials with the exception of Buhl, which is unknown (table 24). The woman at Gordon Creek was tightly flexed, while the individuals at Horn Shelter No. 2 and Wilson-Leonard were semi-flexed.

TABLE 23. ANTHROPOLOGICAL COMPARISON OF SELECTED HUMAN BURIALS WITH UNCALIBRATED RADIOCARBON AGES BETWEEN 9500 AND 10,020 RC YR.

Site	State	Sex	Age at Death	Radiocarbon Date/Age	Cultural Affiliation	Setting	Burial Practice	Ceremonial Aspects	Associated artifacts
Arch Lake	New Mexico	Female	17-21 years	10,020 ± 50 RC yr. BP *11,260 to 11,640 CAL years (1 sigma)*	Uncertain, possibly Plainview	Open air, apparently not within a camp	Primary inhumation, extended	Possible sanctification of interment with red ocher	19 talc beads (necklace); red pigment, resharpened flake tool, probable bone tool (missing, not available for this analysis)
Horn Shelter No. 2	Texas	Female	10 years	9690 ± 50 RC yr. BP	San Patrice (diagnostic artifacts in associated stratum; a Wilson point also found)	Associated with habitation in rock shelter	Double burial, primary inhumation, semi-flexed	Buried with older male. Both individuals placed with heads touching nested turtle shells. Grave prominently marked with layer of 19 rocks that extend from neck toward feet, but leaving heads uncovered (symbolic cara-pace?). Deliberate breakage of needle?	Bone needle (eyed, possibly a few shell beads (17 of 81 beads were recovered in situ, most of these appear associated with the male)
Horn Shelter No. 2	Texas	Male	35-44 years	9710 ± 40 RC yr. BP	San Patrice (diagnostic artifacts found in asso-ciated stratum; a Wilson point also found)	Associated with habitation in rock shelter	Double burial, primary inhumation, semi-flexed	Buried with young female. Many items in flintknapper kit appear recently made (for the burial?). Apparent ties with turtles. Head rests on 3 turtle carapaces. Turtle carapace over face, another beneath the pelvis. Grave prominently marked with layer of 19 rocks that extend from neck toward feet, but leav-ing heads uncovered (symbolic carapace?). Apparent placement of hawk feet and badger claws in/near mouth.	Flintknapper kit (2 abraders, 2 antler billets, early stage biface, large red ocher nodule, elongated cut bone of uncertain function), shaft wrench/flaker, probable necklace with 2 kinds of shell beads and drilled canid teeth, three nested turtle carapaces, 4th carapace found behind pelvis contains 3rd antler tool, fifth partial carapace over face; hawk talons and badger claws in/near mouth strongly suggest animal familiars
Gordon Creek	Colorado	Female	25-30 years	9650 ± 50 RC yr. BP	Hell Gap based on biface reduction tech-nology and radiocarbon date	Open air, apparently not within a camp	Primary inhumation, tightly flexed	Sanctification of interment with abundant red ocher (on grave, body, and artifacts); fire at/near grave and deliberate burning of selected stone tools and animal remains and breakage of selected bone artifacts and animal bone.	Cut small mammal ribs, 4 elk teeth (1 drilled), hammerstone, smooth stone, large, complete biface, three smaller broken bifaces, early stage endscraper, 3 utilized flakes, and ocher. Tool assemblage at early stage of use history.
Wilson-Leonard	Texas	Female burial two	20-25 years	10,000 to 9750 RC yr. (chro-nostratigraphic position)	Wilson (diagnostic artifacts in associated stratum)	Buried on periphery of open air camp	Primary inhumation, semi-flexed	Uncertain. Buried on edge of camp in shallow grave with unmodified fossiliferous limestone slab possibly used to anchor burial covering.	Well-used bifacial mano recycled into bifacial chopper (also used), fossil shark's tooth found near neck.
Buhl	Idaho	Female	17-21 years	10,000–9500 RC yr. (chro-nostratigraphic position)	Windust/ Great Basin Stemmed	Open air; unknown if associated with camp	Largely disturbed, probable primary inhumation	Biface and needle appear newly made (for burial?). Burial disturbed in prehistory and during modern gravel quarrying.	Windust biface (newly made), unmodified badger baculum, incised bone awl or pin, bone needle (eye gouged not rotary drilled)

Sources: Bousman et al. 2002; Breternitz et al. 1971; Green et al. 1998; Guy 1998; Muniz 2004; Owsley et al. this volume; Redder 1985; Redder and Fox 1988; Steele 1998; Sullivan 1998; Young 1988; Young et al. 1987.

TABLE 24. COMPARATIVE DATA ON ARTIFACT AND BODY PLACEMENT AND DEGREE OF DENTAL WEAR.

Site	Body position	Head placement	Face placement	Hand placement	Tooth modification	Artifact placement	Above ground features/ markers
Arch Lake female	Extended burial lying on back	To the southeast	Skyward and tipped slightly to the east	Left forearm flexed at elbow, positioning missing hand across abdomen; right upper arm at side, lower arm not observed	Wear is slight on incisors and first molars.	Beads near neck, red pigment concentrated with flake tool near left elbow, probable bone tool lying diagonally across chest at slightly higher elevation	
Horn Shelter No. 2 female	Double burial, semi-flexed, lying on left side and facing back of adult male	To the south, head bent slightly forward so top of skull touched turtle shells	Facing west (facing adult male and back wall of shelter)	Hands placed in front of face	Heavy wear on the single remaining deciduous second molar due to diet	Needle found in abdominal region between lower ribs and femur.	19 large limestone rocks covered grave from neck toward feet
Horn Shelter No. 2 male	Double burial, semi-flexed, lying on left side	To the south, temporal portion of skull rested on turtle shells	Facing west (facing back wall of rock shelter)	Hands near face	Severe wear, antemortem tooth loss; disproportionate wear of anterior teeth and probable use as tools	Most items found near head; turtle carapace containing antler (billet?) found behind pelvis.	19 large limestone rocks covered grave from neck toward feet
Gordon Creek female	Tightly flexed burial lying on left side, partially displaced	To the north	Uncertain	Uncertain	Disproportionate wear of anterior teeth and probable use as tools	Most artifacts displaced, utilized flake found near left humerus	
Wilson-Leonard female	Semi-flexed lying on her right side	To the northeast	Facing west	Right hand under head, left hand near right wrist	Disproportionate wear of anterior teeth and probable use as tools; antemortem tooth caries on her right side at mid and tooth abscess	Fossil shark tooth at neck; mano/chopper on her right side at mid torso (lying on the left distal femur).	An unmodified slab of fossiliferous limestone may have protruded above ground surface after the shallow grave was filled
Buhl female	Uncertain, skeleton not articulated, and largely displaced	Uncertain	Uncertain	Uncertain	High degree of occlusal wear relative to age of individual; dentine cupping and oblique wear planes reflect a diet containing fine sand or grit	Biface found in situ under cranium, other artifacts displaced	

Sources: Breternitz et al. 1971; Green et al. 1998; Guy 1998; Muniz 2004; Owsley et al. this volume; Powell and Steele 1994; Redder 1985; Redder and Fox 1988; Steele 1998; Young 1988; Young et al. 1987.

TABLE 25. COMPARISON OF BURIAL ASSEMBLAGES BY FUNCTIONAL CATEGORIES AND TYPE OF MODIFICATION.

Site	Number of cultural items*	Ornaments and/or amulets	Emblems of spirit connection**	Fabricating and processing tools	Hunting-related/ weaponry	Unmodified or other raw material	Chipped stone	Ground Stone
Arch Lake female	22 (includes 19 beads)	Necklace	Red pigment?	Flake tool and probable bone tool (missing from collection)	No	Red pigment	Retouched flake tool	No
Horn Shelter No. 2 female	1 (possibly some of the small shell beads)	Possible shell beads	Turtle shells?	Needle, turtle shell?	No		No	No
Horn Shelter No. 2 male	Ca. 102 (includes ca. 85 ornaments, 7 items in flintknapper kit, 5 turtle shells, possible antler tool, 2 hawk feet, badger claw)	Necklace (two kinds of shell and drilled coyote teeth)	Turtle shells, hawk talons, badger claws, coyote teeth, sea shell	Flintknapping kit (antler billets, sandstone abraders, elongated bone blank?), shaft straightener /flaking tool, turtle shells?	No	Large ocher nodule, early stage biface for stone tool making, elongated bone blank	Early stage biface	Sandstone abraders
Gordon Creek female	About 16 (including ocher) *	Elk incisors, drilled	Elk teeth? Group emblem? Gendered?	Hammerstone, smooth stone, bifaces, flake tools	No	Preform, largest biface, red pigment?	4 bifaces 3 flake tools	
Wilson-Leonard female	2–3	Fossil shark tooth	Fossil shark tooth?	Mano/chopper	No	Unmodified fossiliferous limestone manuport	Mano/chopper	Mano/chopper
Buhl female	At least 4 (burial partially destroyed)	Uncertain (burial partially destroyed)	Badger baculum?	Knife, needle	No	Unmodified badger baculum? (needle blank?)	Stemmed biface with chisel-shaped tip	No

* Some broken items not counted if they appear to derive from another specimen; presence of red ocher counted as 1; bones from the right and left foot of a single hawk counted collectively as 2; beads and drilled teeth counted individually.

** Material emblems that connect an individual and/or group to spirit helpers, animal or plant, or mineral families, and/or ancestral totems; can include natural objects embodying personal or group medicine.

Sources: Data from Jodry's examination of Arch Lake and Horn Shelter No. 2 artifact assemblages and information in Breternitz et al. 1971; Green et al. 1998; Guy 1998; Hamlin 2001; Muniz 2004; Redder and Fox 1988; Sullivan 1998.

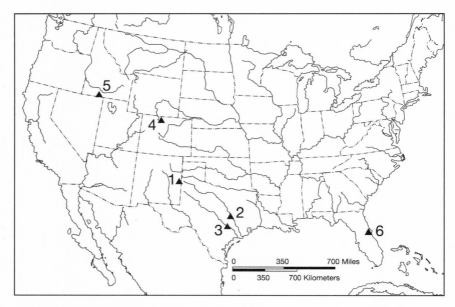

Figure 20. Map showing locations of burials compared in the archaeological analysis: Arch Lake (1), Horn Shelter No. 2 (2), Wilson-Leonard (3), Gordon Creek (4), Buhl (5), and Windover (6).

The sandy sediment at Arch Lake may have facilitated digging a relatively deep burial pit sufficient in size to accommodate a fully extended body. In contrast, the semi-flexed woman at Wilson-Leonard was tightly wedged into a relatively shallow burial pit (Guy 1998:1208).

The long axes of the burials trend roughly north-south with heads variably placed (S, SE, N, and NE). The face of the Arch Lake woman was uniquely oriented skyward, while the three individuals at Wilson-Leonard and Horn Shelter No. 2 faced west. This aspect of burial positioning was unobserved for Gordon Creek and Buhl. The left hand of the Arch Lake woman was positioned across the abdomen; the right lower arm and hand are missing. The hands were placed near the face for the three individuals buried at Wilson-Leonard and Horn Shelter No. 2. This aspect was not observed at Gordon Creek and Buhl.

Burial Goods

Table 25 organizes information on burial goods by functional categories used for analysis of the Windover cemetery (Hamlin 2001). Comparative classes include ornaments, fabricating and processing tools, hunting-related/weaponry, and unmodified or other raw materials. Two additional classes were added for this review: (1) spirit emblems, which connect an individual or group to spirit helpers, and (2) the relative occurrence of chipped versus ground stone implements.

Ornaments

Nearly 150 items were collectively associated with the six paleoburials compared here (table 25). Seventy-four percent of these burial goods were ornaments: 19 talc beads, about 81 shell beads, 8 drilled animal teeth, and a fossil shark's tooth. Arch Lake woman, Horn Shelter No. 2 man, and Wilson-Leonard woman apparently were wearing ornaments and/or amulets around their necks. The woman at Gordon Creek had one elk incisor and three subincisors. The incisor had been biconically drilled through the root; the roots of the subincisors were broken and the presence/absence of perforations is uncertain. The elk teeth were displaced from the grave due to natural erosion and their position relative to the body was not observed. Breternitz and colleagues (1971:178) suggest that these ornaments may have been torn from a necklace or clothing at the time of burial. They further suggest that the historic practice among some Plains groups of adorning female clothing with perforated elk canine teeth (Wood 1957) may have been foreshadowed at Gordon Creek.

Ornaments occurred with at least four of the six individuals (67%) in the Arch Lake comparative sample. In contrast, only five of 145 Windover interments (3.4%) were interred with ornaments consisting of beads made of shell, seeds, antler, or fish vertebrae. At Windover, ornaments were associated with two adult females and three children (Hamlin 2001:128).

Hunting-Related/Weaponry

Hunting-related/weaponry (finished projectile points, atlatl hooks and weights, bola and sling stones, snare triggers) were not found with the ancient burials discussed here. The Windust stemmed biface from Buhl was made (or resharpened) just before interment (Green et al. 1998:450). The intentionally made, chisel-shaped bit on the distal end, rather than a sharp point, provides evidence that this tool was a hafted knife, not a projectile point. A similar chisel-shaped bit is present on the unifacial tool accompanying the Arch Lake interment. The absence of finished projectile points and preforms with the Horn Shelter No. 2 male is notable, as he was interred with a fairly complete flintknapper's tool kit and an early stage biface. As discussed later, Jodry suggests that this individual's role as a hunter was not emphasized within the burial context, but his position as a flintknapper and possible healer is evident.

No finished projectile points were interred at Gordon Creek, although four bifaces, three flake tools, and a hammerstone were among the burial goods. Muniz (2004:261) has identified one of the Gordon Creek bifaces as a late-stage Hell Gap projectile point preform. An alternative suggestion is that this preform was intended for a hafted knife rather than a projectile point. The length: thickness ratio supports this interpretation. The blade thickness falls five standard deviations above the mean for Hell Gap points from the Casper site in Wyoming, while its length is within one standard deviation of the combined mean of points from the Casper and Jones-Miller sites (Muniz 2004:260).

Use-wear analysis indicates that a late-stage Hell Gap preform was used as a butchering knife at the Jones-Miller bison kill (D. Stanford, unpublished notes). Gordon Creek may provide a first glimpse into the nature of tools thought to have been needed by an adult female in a Hell Gap–conceived afterlife.

Hunting-related artifacts and weaponry were associated with less than 10.3% of the Windover documented series (145 individuals). This subset of 2 females, 11 males, and 2 unsexed subadults represents 18.3% of the 82 individuals buried with grave goods. Of 13 sexed individuals with hunting-related artifacts/weaponry, nearly 85% were male (Hamlin 2001:128).

Fabricating and Processing Tools

All six individuals in the ancient sample were interred with tools needed to make other objects and/or to process food and materials. Included in this category are grinding stones, shaft straighteners, awls, scrapers, knives, utilized flakes, choppers, drills, flintknapping tools, and hide-working, bone-cracking, and butchering equipment. Also included are needles, which were classified in the Windover analysis as domestic items, a category subsumed here under fabricating and processing tools.

The Arch Lake woman was interred with a delicate, resharpened flake tool suitable for light duty cutting and scraping and with a robust bone tool. The

Gordon Creek woman was buried with four bifaces, three flake tools, a hammer-stone, and two possible hide-working tools. Eyed bone needles accompanied the females at Horn Shelter No. 2 and Buhl; the latter also had a stemmed biface. The woman at Wilson-Leonard was interred with a heavily used mano that had been recycled into a chopper (Sullivan 1998:705).

At Windover, fabricating and processing tools were more often associated with males (Hamlin 2001). Of the 82 interments with artifacts, 2 of 23 (8.7%) females, 12 of 29 (41.4%) males, and 3 subadults were associated with fabricating and processing tools.

Unmodified or Other Raw Material

This category includes raw materials that "may have been included with the deceased for use in the manufacture of tools or ornaments in the afterlife, or may have been intended as food offerings" (Hamlin 2001:129). Four of the five burials in this study appear to include raw materials and one may include food remains (Gordon Creek).

A pouch of red pigment was likely buried with the Arch Lake woman. Accompanying the Horn Shelter No. 2 male was a large nodule of red ocher, an elongated bone blank from a deer metacarpal, and an early stage biface (material for tool stock). Possible items in this category at Gordon Creek include a large biface (unburned, as were three other bifaces); unshaped, unburned flakes; small mammal ribs that appear to have been "grooved-and-snapped," some red pigment, and possible food remains burned at the grave (Breternitz 1971:179). The Buhl burial included an unmodified badger baculum of uncertain meaning or function. At Wilson-Leonard, a manuport consisting of a large fossiliferous limestone rock was found within the upper burial pit at a slightly higher elevation than the skeleton and mano/chopper (Guy 1998:1210). Although its purpose is unknown, it may have held down a perishable wrap covering the body (Bousman et al. 2002:987).

Chipped versus Ground Stone Tools

In the ancient sample five of six individuals (83%) were accompanied with chipped stone artifacts, the exception being the adolescent female at Horn Shelter No. 2. These artifacts consist of a finished Windust biface from Buhl, a possible Hell Gap preform, a possible Clovis biface, two additional bifaces and three edge-damaged flakes from Gordon Creek, an early stage biface associated with a flintknapper's kit at Horn Shelter No. 2, a bifacial chopper made from a mano at Wilson-Leonard, and a retouched unifacial tool at Arch Lake.

Two individuals (33%) were interred with stone implements used to grind or abrade: a mano at Wilson-Leonard, and two sandstone abraders in a flint-knapper kit with the Horn Shelter No. 2 male. Five individuals (Arch Lake, Gordon Creek, the Horn Shelter No. 2 male and female, and Buhl) were interred with items of stone or bone that were ground and/or abraded during their manufacture (talc beads, elk teeth, and needles).

Although a ground stone tool accompanied the Wilson-Leonard female, only two other ground stone fragments, among 186 other stone tools and nine projectile points, were excavated from the associated Wilson occupations at the site (Bousman et al. 2002:983, 986). Early Archaic contexts at Wilson-Leonard yielded the highest percentage (22%) of ground stone artifacts. The lowest percentage (0.02%) was associated with the early Paleoamerican period. Ground stone implements during the terminal Pleistocene–early Holocene often consisted of abraders used in flintknapping or to grind pigments and ornaments, rather than tools such as manos or pestles used to process plants for food.

Emblems of Spiritual Connection

This category includes amulets, totems, icons, and offerings that connect people with the spirits of particular animals, plants, minerals, and places. These items are thought to have been used, carried, or worn to strengthen relationships that individuals and groups had with entities in the natural and spirit world. These emblems are physical embodiments of and references to these relationships. Emblems are objects that represent something else, usually by suggesting its nature or history, and that are symbolically linked pictorially with their reference. In this analysis, items identified as possible emblems include some of the partial remains of animals and red ocher placed in a way suggestive of ceremonial use at the time of interment.

Human interaction in the spirit world that is based in deep connections with the earth and her creatures is strongly and globally expressed today, particularly among indigenous peoples, who are increasingly sharing this knowledge with people outside their traditions. Emblems of spiritual connection often illustrate important aspects of the history and beliefs of an individual or group (see Fitzhugh and Kaplan 1982). Such emblems have been identified archaeologically for Hell Gap (Stanford 1978) and Folsom (Bement 1999). Boyd (2003) presents a rigorous archaeological study of spirit emblems depicted in Middle Archaic rock art in southwest Texas that is informed by living traditions among the Huichol of Mexico.

Emblems of spiritual connection appear to be associated with at least three of the five ancient burials summarized here: Arch Lake, Horn Shelter No. 2, and Gordon Creek.

The double burial from Horn Shelter No. 2 provides the most compelling evidence for possible emblems of spirit connection. Interred with the male were selected portions of a hawk, badger, turtles, and coyotes. Hawk feet, terminal digits (claws) from a badger paw, and four drilled canine teeth from two coyotes were found in the vicinity of the man's head and neck. At least one hawk foot was located in his mouth. A large portion of a turtle carapace covered his face, and three turtle carapaces lay beneath his head. The head of the adolescent girl was placed in the grave so that it also touched these three turtle shells. Nineteen limestone slabs covered the grave. Their placement was reminiscent of a turtle carapace, in that the rocks extend from the neck toward the feet, leaving the heads uncovered. The hawk, badger, and coyote were associated with the man, while the turtles appear to relate to both individuals. Jodry suggests that the hawk, badger, and coyote were likely part of the man's personal medicine and, as such, were physical embodiments of his spiritual relationships with particular bird and animal familiars. The turtles may speak to a wider group affiliation, such as a family, clan, ethnic, or pan-ethnic totem. Turtles are well represented among food remains identified in the associated habitation levels at both Horn Shelter No. 2 and Wilson-Leonard, highlighting their importance within the regional ecology. Redder and Fox (1988:1) previously recognized the inclusion of both "ritual and utilitarian objects" and "religious paraphernalia" in the Horn Shelter No. 2 interment.

Red ocher distributed over the graves and burial goods of women at Arch Lake and Gordon Creek point toward ceremony conducted at the time of interment. Breternitz et al. (1971) and Muniz (2004) discuss a fire associated with the Gordon Creek interment and intentional breakage and burning of burial objects and animal bone.

It is uncertain whether the Cretaceous-age fossil shark tooth found at the neck of the Wilson-Leonard woman was an emblem or simply an ornament lacking a spiritual connotation. Likewise, the function and meaning of the unmodified badger baculum found with the Buhl burial is uncertain.

Artifact Condition and Placement

Recent disturbance to the burials at Gordon Creek (erosion) and Buhl (gravel quarrying) led to their discovery, but the displaced human remains and burial goods left partial records of these interments.

Items newly or recently manufactured and/or used sparingly were interred at Buhl (needle and biface), Gordon Creek (endscraper and Hell Gap preform), and Horn Shelter No. 2 (abraders, early stage biface, red ocher, turtle carapace beneath the man's hip). Conversely, tools interred at Arch Lake (flake tool) and Wilson-Leonard (mano/chopper) exhibit later stages in their use. The condition of burial goods at Gordon Creek included items thought to have been intentionally broken and/or burned at the time of interment (Breternitz et al. 1971; Muniz 2004). Likewise, the needle found with the girl at Horn Shelter No. 2 had a "green bone" fracture at its tip, indicating breakage before or at the time of interment (Redder and Fox 1988:6).

Most items placed with the man at Horn Shelter No. 2 lay beneath his head (three turtle carapaces, flintknapper kit, ocher nodule, and bone blank). The flintknapping tools, ocher, and blank were tightly clustered and may have been within a perishable bag of some sort. Remnants of a turtle carapace covered his face, and another (newly modified and associated with a possible antler tool) was found under his left hip (Redder 1985 and pers. comm. 2006). A shaft straightener was found in an open area between his radius-ulna and femur. The placement of burial goods with the Buhl woman is largely unknown, with the exception of a newly made biface found beneath her head.

Ornaments or emblems were found near the neck and head at Arch Lake, Horn Shelter No. 2 (male), and Wilson-Leonard, suggesting that they were personal items being worn as necklaces or amulets. The placement of ornaments at Gordon Creek and the presence of ornaments at Buhl were unobserved due to recent disturbance.

Artifact placement may suggest that some items were personal belongings carried in perishable pouches attached to the waist. These include red ocher and a flake tool found near the flexed elbow and lower ribs of the Arch Lake woman and a needle found between the flexed femur and lower ribs of the female at Horn Shelter No. 2.

At Wilson-Leonard, a mano/chopper and an unmodified limestone rock were placed in an open space between the woman's flexed knees and elbows, near the edge of the burial pit opposite her hips (Bousman et al. 2002:987).

Use of the Anterior Teeth as Tools

Dental wear was rapid in the ancient group, reflecting a coarse, abrasive diet and use of the anterior teeth in task activities. Disproportionate wear of the anterior teeth relative to the molars was noted for the females at Wilson-Leonard and Gordon Creek (Powell and Steele 1994). The extent of this wear "approached that seen in Upper Paleolithic hunter and gatherers of Europe and exceeded that reported for Eskimos, a population noted for their use of anterior teeth as tools" (Steele 1998:1452). The ten year old from Horn Shelter No. 2 already shows beginning dentin exposure on her upper and lower incisors; a similar finding with a greater degree of progression was found in Arch Lake. At Buhl the incisors exhibit a moderately advanced degree of wear (stage 6 on a total scale of 8), as do the canines, premolars, and second molars, although less than the first molars (stage 7; Green et al. 1998:447). The Horn Shelter male, aged about forty years, shows complete loss of nearly all tooth crowns, such that function was maintained by occlusion of the root stubs. The anterior teeth show slight buccal-lingual rounding, suggesting use in activities besides normal attrition.

Cultural Affiliation

Five different cultural associations may be represented by the burials compared here (table 23). Only one of the six individuals in this sample was found with a finished diagnostic artifact. This is the Buhl woman interred with a Windust stemmed biface. At Gordon Creek the woman's cultural affiliation has been proposed recently as Hell Gap based on comparative analysis of morphology and lithic reduction strategies evident on a preform for a hafted biface (Muniz 2004).

At Wilson-Leonard and Horn Shelter No. 2 cultural affiliation has been inferred from the chronostratigraphic positions of the burial pits within deep geologic sections and with reference to the dominant projectile point style found in the respective stratum from which each burial is thought to have originated. The top of the burial pit was not detected during the Wilson-Leonard

excavation. Characteristics of sediment comprising the burial pit fill and associated radiocarbon dates indicate that the grave was dug from a surface within
the lower Leanne soil. Nine stemmed Wilson points were recovered from this
soil, which contains at least two superimposed Wilson occupations thought
to represent residential camps (Bousman et al. 2002:983). Also found in these
components were two pieces of ground stone, 186 other chipped stone tools,
pit and rock features, and faunal remains dominated by rabbit, hare, turtle, and
deer (Bousman et al. 2002:988).

Of particular interest here in relation to the Horn Shelter No. 2 burial were
four San Patrice points recovered at the Wilson-Leonard site. One point was
found near the contact of the Leanne soil and the overlying stratum (as was a
displaced Midland point). Three others were found in strata above the Leanne
soil (Dial et al. 1998:397). Bousman et al. (2004) identify this as the middle of
Unit II, which is dated between 8400 and 9500 RC yr. This evidence indicates
that the Wilson-Leonard burial predates the San Patrice occupation at this
locality.

At Horn Shelter No. 2 the top of the burial pit was not detected during
excavation, but the dark gray sediment comprising the burial pit fill confirms
that the pit originated within substratum 5G, not in the overlying strata consisting of distinctive red deposits. Four uncalibrated radiocarbon dates from
the 5G stratum are consistent with this interpretation: 9500 ± 200 (Tx-1830,
charcoal), 9980 ± 370 (Tx-1722, charcoal), $10,030 \pm 150$ (Tx-1998, shell);
$10,310 \pm 150$ (Tx-1997, shell; Redder 1985:41), as are two new AMS human
bone dates from the burial reported here: 9710 ± 40 (CAMS-60681) and $9690
\pm 50$ RC yr. BP (CAMS-51794).

Substratum 5G consists of a very dark midden deposit thought to result
from year-round residential occupation. This stratigraphic unit contains the
densest cultural deposits in the paleo levels of the shelter (Redder 1985:41,
47). Animal bones recovered while screening the burial fill include turtle, deer,
rabbit, bird, rodent, snake, frog, and four species of fish (Redder 1985:43; Redder and Fox 1988:9). Projectile points found in substratum 5G and underlying
stratum 5F include six San Patrice (Brazos Fishtailed) points and a possible
Wilson point. The latter consists of two conjoined pieces. The base portion
was recovered in substratum 5G and the midsection in underlying stratum 5F.
Another Wilson point was recovered from the north end of Horn Shelter No. 2

in paleo deposits excavated by Forrester (1985, fig. 2F). Current evidence in-dicates that the double burial at Horn Shelter No. 2 was affiliated with a San Patrice (Brazos Fishtailed) occupation and begs the question of whether San Patrice points overlapped in time with Wilson points at this locality.

The remaining burial is Arch Lake, for which cultural affiliation is uncer-tain. The AMS date of 10,020 ± 50 RC yr. BP overlaps radiometric determina-tions of about 10,000 yr. BP for Plainview at the nearby sites of Lubbock Lake and Lake Theo in Texas (table 26), where Plainview occurs stratigraphically above Folsom (Holliday 2000:268; Knudson, Johnson and Holliday 1998). Also contemporaneous with Arch Lake are constricting stem lanceolate points including Agate Basin and Hell Gap (table 26). Lubbock is another point type that temporally overlaps the Arch Lake burial. Lubbock points are contempo-raneous with Plainview at the Lubbock Lake site, both stratigraphically and radiometrically, dating to 9950 ± 120 RC yr. BP (Holliday 2000:269). Kerr and Dial (1998:473) suggest a possible age range for Lubbock points between 9900 and 10,300 yr. BP. Holliday suggests that Lubbock Points may be Southern Plains equivalents of the Agate Basin point type, while Kerr and Dial note similarities with Plainview. Another lanceolate point type from the Llano Es-tacado region is Milnesand.

Sorting out typological relationships and relative temporal positions among these late Paleo point types that may have been contemporaries, circa 10,000 yr. BP, is the tip of the iceberg relative to deducing what this all means regarding the cultural affiliation of a particular individual, such as the woman interred at Arch Lake.

Plainview/Milnesand/Lubbock points appear to be more common in the vi-cinity of Arch Lake than seems to be the case for either Agate Basin or Hell Gap. A large Plainview camp, the Warnica-Wilson site, is located about fifteen miles northwest of Portales and roughly thirty-two miles from the Arch Lake locality. This surface assemblage is thought to consist primarily of Plainview materials deposited over multiple visits. The presence of numerous thick endscrapers that were heavily used and resharpened suggests that hide working was frequently undertaken there (Reutter 1996). Of note was the recovery of several biconi-cally drilled beads made of stone. Warnica notes that the beads were larger and different in appearance than the talc beads at Arch Lake, although they do sug-gest that Plainview folks fashioned biconically drilled stone beads.

TABLE 26. RADIOCARBON DATES FOR PROJECTILE POINT STYLES CONTEMPORANEOUS WITH THE ARCH LAKE BURIAL (10,020 ± 50 RC YR. BP).

Point style	¹⁴C AGE (RC yr. BP)	Lab No.	Archaeological site	Material dated	Reference
Plainview	9950 ± 110	SMU-866	Lake Theo, Tex.	Soil, humic acid	Johnson, Holliday, and Neck 1982
Plainview	10,015 ± 80	SI-3203	Lubbock Lake, Tex.	Organic rich mud, humin	Holliday et al. 1983
Plainview	9990 ± 100	SMU-728	Lubbock Lake, Tex.	Organic rich mud, humic acid	Holliday and Johnson 1981
Plainview	9960 ± 80	SMU-275	Lubbock Lake, Tex.	Humic acid	Johnson and Holliday 1980
Plainview	10,090 ± 100	Average of TX-153, 658, 657, and 346	Bonfire Shelter, Tex.	All on charcoal	Haynes, pers. comm. 1998, in Holliday 2000: tableVIB
Agate Basin	9860 ± 200	Average of M-1131 and 0-1252	Agate Basin, Wyo. (Brewster)	Both on charcoal	Calculation by D. Meltzer, 1999 (after Hietala 1989)
Hell Gap	10,020 ± 320	SI-1989	Jones-Miller, Colo.	Charcoal	Bonnichsen et al. 1987
Hell Gap	9830 ± 350	RL-125	Casper, Wyo.	Charcoal	Frison 1974
Hell Gap	10,060 ± 170	RL-208	Casper, Wyo.	Bone	Frison 1974

Sources: Plainview dates taken from Holliday 2000: table VIB. Reliable radiocarbon ages for Plainview assemblages.

Hell Gap burial practice, as inferred from Gordon Creek, appears quite different from that observed at Arch Lake. At Gordon Creek the woman was tightly flexed on her side and accompanied by a relatively large burial assemblage that received ceremonial treatment in the form of abundant red ocher, the breakage and burning of tools, and presence of food remains. In contrast, the woman at Arch Lake was extended on her back and interred with a relatively small burial assemblage that is unbroken and unburned and includes tools that appear to have been farther along in their use-lives than those found at Gordon Creek. The woman at Arch Lake appears to have been buried with a few personal belongings, most of which she may have been wearing. One might envision a deliberately planned aspect to the burial undertaken at Gordon Creek versus a more impromptu aspect to the interment at Arch Lake.

The greatest similarities in burial practice observed in this study involve the Horn Shelter No. 2 and Wilson-Leonard burials, which are closest to one another geographically and perhaps temporally. In both cases the burials were associated with residential camps, and body position was semi-flexed with the long axis oriented roughly north-to-south and the heads facing west. Neither interment included the application of red ocher over the grave or body. The females at both sites were accompanied by few nonperishable burial goods.

These ancient individuals present precious opportunities to learn about Paleo lifeways. We have much to learn about variability that existed in burial practices, either within a cultural group or among contemporaneous groups.

Discussion and Conclusion

✦

HE Arch Lake woman is among the oldest Paleoamerican human remains yet found. The remains of this young woman were buried fully
extended in a grave that was dug 1.1 m (3 feet 6 inches) deep into
a sandy deposit. The cross section of the pit, sketched and photographed in
1967, indicate that the grave walls were nearly vertical. Associated materials
included a probable bone tool, a resharpened flake tool, a type of red ocher,
and talc beads. Talc is a white to dark green hydrous magnesium silicate. It is a
secondary alteration mineral usually found in metamorphic rocks. Because of
its soft, smooth texture it was commonly used in native carvings, although this
is the oldest example from a dated North American burial context.

The Arch Lake skeleton was dated by different radiocarbon laboratories
using multiple chemical fractions to determine the remains' geologic age and
paleodietary information. Because the skeleton had been treated with a variety
of preservatives that required sequential removal and the collagen preservation was variable among skeletal elements, laboratory results varied as laboratory work proceeded. The best estimate of the skeleton's age is $10,020 \pm 50$ RC
yr. BP (CAMS-61133), which corresponds to a 1-sigma calibrated age range
of 11,260 to 11,640 CAL yr. BP (table 4). The best estimates for the stable isotope values from bone collagen are $\partial^{13}C = -14.1‰$ (PDB) and $\partial^{15}N = +13.0‰$
(AIR). These data are consistent with an omnivorous diet containing significant amounts of animal protein.

Cultural affiliation of the remains is undetermined, neither the talc beads
nor the stone tool being culturally diagnostic. Evidence for biological affiliation was garnered through analyses of the cranium, dentition, and post-cranial
skeleton, and shows strong differentiation in all aspects from modern Native
Americans. In addition, the Molecular Anthropology Laboratory of the University of California at Davis processed a small sample of the right femur in an
attempt to recover mitochondrial DNA; none was obtained.

Cranial and post-cranial morphology show the Arch Lake woman to be different from other ancient American skeletons in some respects, similar in others, and strongly differentiated from modern Native Americans. Cranial vault shape differs from all other early crania in being brachycranic, although Gordon Creek, characterized as mesocranic, approaches this condition. The Arch Lake cranium is similar to other early American crania in having a low face. The low face appears to be the most consistent feature of early Americans and is also found in early people of the eastern Pacific Rim (Jantz and Owsley 2005).

The general conclusions to be reached from Arch Lake woman's post-cranial skeleton is that she was tall, evidently taller than any other female skeleton from the period 8000 RC years ago or earlier. Some of these early individuals, such as from Gordon Creek, are quite small (Breternitz et al. 1971). Robusticity varies considerably depending upon anatomical region. The femur midshaft and tibia are not especially robust. The subtrochanteric femur is more robust, and the humerus is very robust. Humerus and femur muscle attachments are not strongly developed, although this is at least partially due to her young age. Wescott's (2001) results for the femur could be taken to mean that the Arch Lake woman was active in the sense of placing loads on the legs, rather than covering long distances on foot. Humerus robusticity suggests a pattern of activity emphasizing upper body strength. This, in combination with a diet including significant quantities of meat implied by a $\delta^{15}N$ bone collagen value of +13.0‰, may suggest that bison hide working was part of this pattern of activity.

The absence of femur platymeria differentiates the Arch Lake skeleton from modern Native Americans, especially those of the Southwest, where it is marked. The overall morphological pattern distinguishes Arch Lake from modern Native Americans and most early Americans. It therefore supports previous analyses showing that early Americans are distinct from recent Indians, while at the same time increasing the apparent diversity seen among early Americans.

The youth and relatively unworn dentition of the Arch Lake woman facilitate analysis of dental features as a means of assessing population relationships. Christy Turner has conducted extensive background work in this regard, noting a classification distinction between two Asian groups: northern

"Sinodonts," including New World populations even in antiquity, vs. southern "Sundadonts." Turner (1987, 1990, 1992) points out that sundadonty is characterized by a retained and somewhat simplified dental pattern, and sinodonty has a pattern of trait intensification so far unknown to have occurred before 15,000 years ago. Interestingly, the dental morphology of the two early Holocene Paleoamericans studied here, Arch Lake and the Horn Shelter No. 2 juvenile, is more consistent with a sundadont definition and less like the sinodont pattern of more recent Native Americans. This correspondence does not necessarily indicate that the early inhabitants of the New World originated in a Southeast Asian–like sundadont population. There might have been a contribution from central and/or western Asians who arrived in East Asia via a southern Siberian route (Lahr 1996; Underhill et al. 2001; Uinuk-Ool et al. 2003). Regardless of the debate for the peopling of Northeast Asia and the homelands of Native Americans, the present findings suggest that the dental morphological characteristics of some early inhabitants in the Americas differ from those of recent Native Americans. Differences include less pronounced shovel-shaping of the maxillary central incisors and the absence of winging and double shoveling. The central upper incisors of the Wilson-Leonard cranium have also been described as being less shoveled than northern Asians and North American Indians (Phelps et al. 1994). Likewise uncommon in Native Americans is the presence of a cusp of Carabelli in Horn Shelter No. 2, and this trait is also present in an eight- to ten-year-old child dated to 9470 RC yr. BP from the Grimes Point Burial Shelter (26CH1C, NSM-743) in the Carson Sink area of Nevada (Tuohy and Dansie 1997).

Farsightedness of the discovery team protected the remains, which otherwise would have been lost. Their recovery notes and effort to preserve the skeleton enabled the present study, allowing us in 2000 to examine these ancient remains as if we were part of the original team. The result was an interdisciplinary investigation that included additional work at the site and analysis of the skeleton and associated artifacts. Diverse specialists with enhanced analytical capabilities and access to comparative databases contributed to the results summarized in this presentation. Many of the analytical methods used were not available four decades ago.

References

Antevs, Ernst. 1955. Geological-climatic dating in the West. *American Antiquity* 20(4): 317–35.

Bement, Leland C. 1999. *Bison Hunting at Cooper Site: Where Lightning Bolts Drew Thundering Herds.* University of Oklahoma Press, Norman.

Bocherens, Hervé. 2000. Preservation of isotopic signals (^{13}C, ^{15}N) in Pleistocene mammals. In *Biogeochemical Approaches to Paleodietary Analysis,* edited by Stanley H. Ambrose and M. Anne Katzenberg, 65–88. Kluwer Academic–Plenum Publishers, New York.

Bonnichsen, Robson, Dennis Stanford, and James L. Fastook. 1987. Environmental change and developmental history of human adaptive patterns: The Paleoindian case. In *North America and Adjacent Oceans during the Last Deglaciation,* edited by William F. Ruddiman and Herbert E. Wright, 403–24. Geology of North America K-3. Geological Society of America, Boulder, Colo.

Bousman, C. Britt, Barry W. Baker, and Anne C. Kerr. 2004. Paleoindian archeology in Texas. In *The Prehistory of Texas,* edited by Timothy K. Perttula, 15–97. Texas A&M University Press, College Station.

Bousman, C. Britt, Michael B. Collins, Paul Goldberg, Thomas Stafford, Jan Guy, Barry W. Baker, D. Gentry Steele, Marvin Kay, Anne Kerr, Glen Fredlund, Phil Dering, Vance Holliday, Diane Wilson, Wulf Gose, Susan Dial, Paul Takac, Robin Balinsky, Marilyn Masson, and Joseph F. Powell. 2002. The Paleoindian-Archaic transition in North America: New evidence from Texas. *Antiquity* 76:980–90.

Boyd, Carolyn E. 2003. *Rock Art of the Lower Pecos.* Texas A&M University Press, College Station.

Breternitz, David A., Alan C. Swedlund, and Duane C. Anderson. 1971. An early burial from Gordon Creek, Colorado. *American Antiquity* 36(2):170–82.

Chidester, Alfred H., and H. W. Worthington. 1962. *Talc and Soapstone in the United States.* Mineral Investigations Resource Map MR-31. U.S. Geological Survey, Washington, D.C.

Cole, Theodore M. III. 1994. Size and shape of the femur and tibia in Northern Plains Indians. In *Skeletal Biology in the Great Plains: Migration, Warfare, Health and Subsistence,* edited by Douglas W. Owsley and Richard L. Jantz, 219–33. Smithsonian Institution Press, Washington, D.C.

Collier, Stephen. 1989. The influence of economic behavior and environment upon robusticity of the post-cranial skeleton: A comparison of Australian Aborigines and other populations. *Archaeology in Oceania* 24(1):17–30.

Dana, Edward Salisbury. 1932. *A Text-Book of Mineralogy, with an Extended Treatise on Crystallography and Physical Mineralogy.* 4th edition. Revised by William E. Ford. John Wiley and Sons, New York.

Defrise-Gussenhoven, E. 1967. Generalized distance in genetic studies. *Acta Geneticae Medicae et Gemellologiae* 17:275–88.

Dial, Susan W., Anne C. Kerr, and Michael B. Collins. 1998. Projectile points. In *Wilson-Leonard: An 11,000-year Archeological Record of Hunter-Gatherers in Central Texas,* vol. 2: *Chipped Stone Artifacts,* assembled and edited by Michael B. Collins, pp. 313–445. Studies in Archeology 31, Texas Archeological Research Laboratory, University of Texas at Austin, and Report 10, Archeology Studies Program, Texas Department of Transportation, Environmental Affairs Division, Austin.

Dobberstein, R. C., M. J. Collins, O. E. Craig., G. Taylor, K. E. H. Penkman, and S. Ritz-Timme. 2009. Archaeological collagen: Why worry about collagen diagenesis? *Archaeological and Anthropological Science* 1:31–42.

Faulhaber, Johanna. 1970. Anthropometry of living Indians. In *Handbook of Middle American Indians,* vol. 9: *Physical Anthropology,* edited by Robert Wauchope and T. Dale Stewart, 82–104. University of Texas Press, Austin.

Fitzhugh, William W., and Susan A. Kaplan. 1982. *Inua: Spirit World of the Bering Sea Eskimo.* Smithsonian Institution Press, Washington, D.C.

Forrester, R. E. 1985. Horn Shelter Number 2: The north end, a preliminary report. *Central Texas Archeologist* 10:21–35.

Frison, George C. 1974. *The Casper Site: A Hell Gap Bison Kill on the High Plains.* Academic Press, New York.

Gill, George W. 1995. Challenge on the frontier: Discerning American Indians from whites osteologically. *Journal of Forensic Science* 40(5):783–88.

Greene, Robert C. 1995. Talc Resources of the Conterminous United States. Open-File Report OF 95–586. U.S. Department of the Interior, U.S. Geological Survey, Menlo Park, Calif.

Green, Thomas J., Bruce Cochran, Todd W. Fenton, James C. Woods, Gene L. Titmus, Larry Tieszen, Mary Anne Davis, and Susanne J. Miller. 1998 The Buhl burial: A Paleoindian woman from southern Idaho. *American Antiquity* 63(3):437–56.

Guy, Jan. 1998. Analysis of cultural and natural features. In *Wilson-Leonard: An 11,000-year Archeological Record of Hunter-Gatherers in Central Texas,* vol. 4: *Archeological Features and Technical Analyses,* assembled and edited by Michael B. Collins, pp. 1067–212. Studies in Archeology 31, Texas Archeological Research Laboratory, University of Texas at Austin, and Report 10, Archeology Studies Program, Texas Department of Transportation, Environmental Affairs Division, Austin.

Hall, Roberta, Diana Roy, and David Boling. 2004. Pleistocene migration routes into the Americas: Human biological adaptations and environmental constraints. *Evolutionary Anthropology* 13(4):132–44.

Hamlin, Christine. 2001. Sharing the load: Gender and task division at the Windover archaeological site. In *Gender and the Archaeology of Death,* edited by Bettina Arnold and Nancy L. Wicker, pp. 119–35. AltaMira Press, Walnut Creek, Calif.

Haynes, C. Vance Jr. 1968. Geochronology of late-Quaternary alluvium. In *Means of Correlation of Quaternary Successions,* edited by Roger B. Morrison and Herbert E. Wright Jr., 591–631. University of Utah Press, Salt Lake City.

———. 1991. Geoarchaeological and paleohydrological evidence for a Clovis-age drought in North America and its bearing on extinction. *Quaternary Research* 35:438–50.

———. 1995. Geochronology of paleoenvironmental change: Clovis type site, Blackwater Draw, New Mexico. *Geoarchaeology* 10(5):317–88.

Herrmann, Nicholas P., Richard L. Jantz, and Douglas W. Owsley. 2006. Buhl revisited: 3-D photographic reconstruction and morphometric re-evaluation. In *El Hombre Temprano en América y sus Implicaciones en el Poblamiento de la Cuenca de México,* edited by Sylvia González and J. C. Jiménez. INAH Publication, Mexico.

Hietala, Harold. 1989. Contemporaneity and occupation duration of the Kubbaniyan sites: An analysis and interpretation of the radiocarbon record. In *The Prehistory of Wadi Kubbaniya* , vol. 2, series assembled by Fred Wendorf and Romuald Schild, edited by Angela E. Close, 284–91. Southern Methodist University Press, Dallas.

Holliday, Trenton W. 1999. Brachial and crural indices of European late Upper Paleolithic and Mesolithic humans. *Journal of Human Evolution* 36:549–66.

Holliday, Vance T. 1995. *Stratigraphy and Paleoenvironments of Late Quaternary Valley Fills on the Southern High Plains.* Geological Society of America Memoir 186. Boulder, Colo.

———. 2000. The evolution of Paleoindian geochronology and typology on the Great Plains. *Geoarchaeology: An International Journal* 15(3):227–90.

Holliday, Vance T., and Eileen Johnson. 1981. An update on the Plainview occupation at the Lubbock Lake site. *Plains Anthropologist* 26:251–53.

Holliday, Vance T., Eileen Johnson, Herbert Haas, and Robert Stuckenrath. 1983. Radiocarbon ages from the Lubbock Lake site, 1950–1980: Framework for cultural and ecological change on the Southern Plains. *Plains Anthropologist* 28(101):165–82.

Jantz, Lee Meadows, Richard L. Jantz, Nicholas P. Herrmann, Corey S. Sparks, Katherine E. Weisensee, and Derinna V. Kopp. 2002. Archaeological Investigations at the Last Spanish Colonial Mission established on the Texas Frontier: Nuestra Señora del Refugio (41RF1), Refugio County, Texas. Archaeological Survey Report no. 315. Center for Archaeological Research, University of Texas at San Antonio.

Jantz, Richard L., and Douglas W. Owsley. 1997. Pathology, taphonomy, and cranial morphometrics of the Spirit Cave mummy. *Nevada Historical Society Quarterly* 40(1): 62–84.

———. 2001. Variation among early North American crania. *American Journal of Physical Anthropology* 114(2):146–55.

———. 2005. Circumpacific populations and the peopling of the New World: Evidence from cranial morphometrics. In *Paleoamerican Origins: Beyond Clovis,* edited by Robson Bonnichsen, 267–76. Texas A&M Press, College Station.

Johnson, Eileen, and Vance T. Holliday. 1980. Plainview kill/butchering locale on the Llano Estacado: The Lubbock Lake site. *Plains Anthropologist* 25(88):89–111.

Johnson, Eileen, Vance T. Holliday, and Raymond W. Neck. 1982. Lake Theo: Late Quaternary paleoenvironmental data and new Plainview (Paleoindian) date. *North American Archaeologist* 3:113–37.

Jolicoeur, Pierre. 1999. *Introduction to Biometry.* Kluwer Academic–Plenum Publishers, New York.

Kerr, Anne C., and Susan W. Dial. 1998. Statistical analysis of unfluted lanceolate and early bifurcate stem projectile points. In *Wilson-Leonard: An 11,000-year Archeological Record of Hunter-Gatherers in Central Texas,* vol. 2: *Chipped Stone Artifacts,* assembled and edited by Michael B. Collins, 447–505. Studies in Archeology 31, Texas Archeological Research Laboratory, University of Texas at Austin, and Report 10, Archeology Studies Program, Texas Department of Transportation, Environmental Affairs Division, Austin.

Klein, Cornelis, and Cornelius S. Hurlbut Jr. 1999. *Manual of Mineralogy* (after James D. Dana). 21st edition, revised. John Wiley and Sons, New York.

Klug, Harold P., and Leroy E. Alexander. 1974. *X-Ray Diffraction Procedures for Polycrystalline and Amorphous Materials.* 2nd edition. John Wiley and Sons, New York.

Konigsberg, Lyle W., Samantha M. Hens, Lee Meadows Jantz, and William L. Jungers. 1998. Stature estimation and calibration: Bayesian and maximum likelihood perspective in physical anthropology. *Yearbook of Physical Anthropology* 41:65–92.

Knudson, Ruthann, Eileen Johnson, and Vance T. Holliday. 1998. The 10,000-year-old Lubbock artifact assemblage. *Plains Anthropologist* 43:239–56.

Lahr, Marta Mirazon. 1996. *The Evolution of Modern Human Diversity: A Study of Cranial Variation.* Cambridge University Press, Cambridge.

Larsen, Clark Spencer. 1997. *Bioarchaeology: Interpreting Behavior from the Human Skeleton.* Cambridge University Press, New York.

Malde, Harold E., and Asher P. Schick. 1964. Thorne Cave, Northeastern Utah: Geology. *American Antiquity* 30(1):60–73.

Moorrees, Coenraad F. A., Elizabeth A. Fanning, and Edward E. Hunt Jr. 1963. Age variation of formation stages for ten permanent teeth. *Journal of Dental Research* 42(6):1490–1502.

Muhs, Daniel R., Thomas W. Stafford, Scott D. Cowherd, Shannon A. Mahan, R. Kihl, Paula B. Maat, Charles A. Bush, and J. Nehring. 1996. Origin of the quaternary dune fields of northeastern Colorado. *Geomorphology* 17:129–49.

Muniz, Mark P. 2004. Exploring technological organization and burial practices at the Paleoindian Gordon Creek Site (5LR99). *Plains Anthropologist* 49(191):253–79.

Phelps, Elisa, Joan Few, Betty Pat Gatliff, D. Gentry Steele, and Frank A. Weir. 1994 (for 1991). Burial to bronze: Excavation, analysis, and facial reconstruction of a burial from the Wilson-Leonard Site, Texas. *Bulletin of the Texas Archaeological Society* 62:75–86.

Powell, Joseph F., and D. Gentry Steele. 1994. Diet and health of Paleoindians: An examination of early Holocene human dental remains. In *Paleonutrition: The Diet and Health of Prehistoric Americans,* edited by Kristin D. Sobolik, 178–94. Occasional Paper no. 22. Center for Archaeological Investigations, Southern Illinois University, Carbondale.

Redder, Albert J. 1985. Horn Shelter Number 2: The south end, a preliminary report. *Central Texas Archeologist* 10:37–65.

Redder, Albert J. and John W. Fox. 1988. Excavation and positioning of the Horn Shelter's burial and grave goods. *Central Texas Archeologist* 11:2–15.

Reutter, Stacie. 1996. A Technological and Typological Study of a Plainview Site on the High Plains of Eastern New Mexico. Master's thesis, Department of Anthropology, Eastern New Mexico University, Portales.

Ruff, Christopher B. 2000. Biomechanical analyses of archaeological human skeletons. In *Biological Anthropology of the Human Skeleton,* edited by M. Anne Katzenberg and Shelley R. Saunders, 71–102. Wiley-Liss, New York.

Scott, Glenn R. 1963. Quaternary geology and geomorphic history of the Kassler Quadrangle, Colorado. U.S. Geological Survey Prof. Paper 421-A. Submitted to the U.S. Department of the Interior, Washington, D.C.

Scott, G. Richard, and Christy G. Turner II. 1997. *The Anthropology of Modern Human Teeth: Dental Morphology and Its Variation in Recent Human Populations.* Cambridge University Press, Cambridge.

Spence, Hugh S. 1940. *Talc, Steatite, and Soapstone; Pyrophyllite.* Publication no. 803. Canadian Department of Mines and Resources, Bureau of Mines, Ottawa.

Stafford, Thomas W. Jr. 1998. Radiocarbon chronostratigraphy. In *Wilson-Leonard: An 11,000-year Archeological Record of Hunter-Gatherers in Central Texas,* vol. 4: *Archeological Features and Technical Analysis,* assembled and edited by Michael B. Collins, 1039–66. Studies in Archeology 31, Texas Archeological Research Laboratory, University of Texas at Austin, and Report 10, Archeology Studies Program, Texas Department of Transportation, Austin.

Stafford, Thomas W. Jr., P. E. Hare, Lloyd Currie, A. J. T. Jull, and Douglas J. Donahue. 1991. Accelerator radiocarbon dating at the molecular level. *Journal of Archaeological Science* 18(1):35–72.

Stafford, Thomas W. Jr., Holmes A. Semken Jr., Russell Wm. Graham, Walter F. Klippel, Anastasia Markova, Nikolai G. Smirnov, and John Southon. 1999. First AMS ^{14}C dates documenting contemporaneity of non-analog species in late Pleistocene mammal communities. *Geology* 27(10):903–906.

Stanford, Dennis, J. 1978. The Jones-Miller site: An example of Hell Gap Bison procurement strategy. In *Bison Procurement and Utilization: A Symposium,* edited by Leslie B. Davis and Michael Wilson, 90–97. Memoir 14. Society for American Archaeology.

Steele, D. Gentry. 1998. Human biological remains. In *Wilson-Leonard: An 11,000-year Archeological Record of Hunter-Gatherers in Central Texas,* vol. 5: *Special Studies,* assembled and edited by Michael B. Collins, 1441–58. Studies in Archeology 31, Texas Archeological Research Laboratory, University of Texas at Austin, and Report 10, Archeology Studies Program, Texas Department of Transportation, Environmental Affairs Division, Austin.

Steele, D. Gentry, and Joseph F. Powell. 1992. Peopling of the Americas: Paleobiological evidence. *Human Biology* 64:303–36.

————. 1994. Paleobiological evidence of the peopling of the Americas: A morphometric view. In *Method and Theory for Investigating the Peopling of the Americas*, edited by Robson Bonnichsen and D. Gentry Steele, 141–63. People of the Americas Publications. Center for the Study of the First Americans, Oregon State University, Corvallis.

Sullivan, Lauren A. 1998. Ground and other nonchipped stone artifacts. In *Wilson-Leonard: An 11,000-year Archeological Record of Hunter-Gatherers in Central Texas*, vol. 3: *Artifacts and Special Artifact Studies*, assembled and edited by Michael B. Collins, 703–21. Studies in Archeology 31, Texas Archeological Research Laboratory, University of Texas at Austin, and Report 10, Archeology Studies Program, Texas Department of Transportation, Environmental Affairs Division, Austin.

Tuohy, Donald R., and Amy J. Dansie. 1997. New information regarding early Holocene manifestations in the western Great Basin. *Nevada Historical Society Quarterly* 40(1): 24–53.

Trotter, Mildred, and Goldine C. Gleser . 1952. Estimation of stature from long bones of American Whites and Negroes. *American Journal of Physical Anthropology* 10(4):463–514.

Turner, Christy G. II. 1984. Advances in the dental search for Native American origins. *Acta Anthropogenetica* 8:23–78.

————. 1987. Late Pleistocene and Holocene population history of East Asia based on dental variation. *American Journal of Physical Anthropology* 73(3):305–21.

————. 1990. Major features of sundadonty and sinodonty, including suggestions about East Asian microevolution, population history, and late Pleistocene relationships with Australian Aboriginals. *American Journal of Physical Anthropology* 82(3):295–317.

————. 1992. Microevolution of East Asian and European populations: A dental perspective. In *The Evolution and Dispersal of Modern Humans in Asia*, edited by Takeru Akazawa, Kenichi Aoki, and Tasuku Kimura, 415–38. Hokusen-Sha, Tokyo.

Turner, Christy G. II, Christian R. Nichol, and G. Richard Scott. 1991. Scoring procedures for key morphological traits of the permanent dentition: The Arizona State University dental anthropology system. In *Advances in Dental Anthropology*, edited by Marc A. Kelley and Clark Spencer Larsen, 13–31. Wiley-Liss, New York.

Uinuk-Ool, Tatiana S., Naoko Takezaki, and Jan Klein. 2003. Ancestry and kinships of Native Siberian populations: The HLA evidence. *Evolutionary Anthropology* 12(5):231–45.

Underhill, Peter A., Giuseppe Passarino, Alice A. Lin, Peidong Shen, Marta Mirazon Lahr, Robert A. Foley, Peter J. Oefner, and L. Luca Cavalli-Sforza. 2001. The phylogeography of Y chromosome binary haplotypes and the origins of modern human populations. *Annals of Human Genetics* 65:43–62.

Weisensee, Katherine Elizabeth. 2001. Pecos Revisited: A Modern Analysis of Earnest Hooton's *The Indians of Pecos Pueblo*. Master's thesis, University of Tennessee, Knoxville.

Wescott, Daniel J. 2001. *Structural Variation in the Humerus and Femur in the American Great Plains and Adjacent Regions: Differences in Subsistence Strategy and Physical Terrain.* Ph.D. dissertation, University of Tennessee, Knoxville.

Wood, W. Raymond. 1957. Perforated elk teeth: A functional and historical analysis. *American Antiquity* 22(4):381–87.

Young, Diane E., Suzanne Patrick, and D. Gentry Steele. 1987. An analysis of the Paleoindian double burial from Horn Shelter No. 2, in Central Texas. *Plains Anthropologist* 32(117):275–98.

Young, Diane E. 1988. The double burial at Horn Shelter: An osteological analysis. *Central Texas Archeologist* 11:11–115.

Index

Page numbers in *italic* type indicate illustrations, tables and figures; page numbers in **bold** type indicate maps.

Topics refer to Arch Lake burial site unless otherwise noted.

Passim = throughout the page range, but noncontiguous.